BOHM–BIEDERMAN CORRESPONDENCE

BOHM–BIEDERMAN CORRESPONDENCE

DAVID BOHM AND CHARLES BIEDERMAN

Volume One:

Creativity and Science

Edited by

Paavo Pylkkänen

London and New York

First published 1999
by Routledge
11 New Fetter Lane, London EC4P 4EE

Simultaneously published in the USA and Canada
by Routledge
29 West 35th Street, New York, NY 10001

Typeset in Baskerville by Keystroke, Jacaranda Lodge, Wolverhampton
Printed and bound in Great Britain by TJ International Ltd, Padstow, Cornwall

British Library Cataloguing in Publication Data
A catalogue record for this book is available from the British Library

Library of Congress Cataloging in Publication Data
Bohm, David.
Bohm–Biederman correspondence / David Bohm and Charles Biederman; edited by Paavo Pylkkänen.
p. cm.
Includes bibliographical references.
Contents: v. 1. Creativity and science –
1. Bohm, David–Correspondence. 2. Physicists–England–Correspondence. 3. Biederman, Charles Joseph, 1906– – Correspondence. 4. Artists–United States–Correspondence. 5. Art and science. 6. Creation (Literary, artistic, etc.) 7. Creative ability in science. I. Biederman, Charles Joseph, 1906– .
II. Pylkkänen, P. (Paavo) III. Title.
N72.S3B64 1999
001'.092'2–dc21
[B] 98–24759
CIP

ISBN 0–415–16225–4

CONTENTS

ACKNOWLEDGEMENTS

The publication of the Bohm–Biederman correspondence has been a large project involving the generous help of many people. We are, of course, extremely lucky that these letters have been preserved. This is thanks to Charles Biederman, who preserved copies of the correspondence. A very important role has also been played by David Bohm's wife Sarah, who has been energetic in her endeavour to further Bohm's work in many areas. She initiated this project by contacting Biederman and has been responsible for many of the various arrangements required to manage the vast project, which the publication of a 4,000-page correspondence inevitably is. It has been a great joy to work with Sarah.

Very special thanks also go to the English artist Anthony Hill, who not only provided me with a lot of material on and by Biederman, but also shared his fascinating views on Biederman's approach and its relation to Bohm's thinking. Arleta Griffor has done an invaluable job in organizing the material and in keying Bohm's hand-written letters onto diskette. Her deep and extensive knowledge of Bohm's philosophy has been very helpful throughout the project. Thanks are also due to Sampo Karjalainen, who scanned Biederman's typed letters. Professor Basil Hiley, Lee Nichol and David Peat have contributed to the project in various ways, not least by discussing many of the ideas.

This project has been supported financially by the Academy of Finland, the University of Skövde, Sweden, and by the Swedish Herbal Institute Ltd, Gothenburg, Sweden. I am especially grateful to the head of the Department of Humanities at Skövde, Seppo Luoma-Keturi, as well as to the Director of the Swedish Herbal Institute, Georg Wikman, for their patronage. I should finally warmly thank Adrian Driscoll and the editorial staff at Routledge who have recognized the special value of this project and helped us in many ways.

Paavo Pylkkänen

FOREWORD

I met David Bohm in 1943 when I began, with other physicists, to work on the horrendous atom bomb Manhattan Project at Berkeley California. Bohm was very highly regarded as a physicist, but what specially interested me was his exceptional breadth of interests which ranged over the whole of human culture. This unusual breadth linked with the fact that Quantum Theory, which was his special area of research, involved revolutionary breadth of thinking. Bohm was specially interested in the philosophical challenge of the new Theory (which Einstein had failed to deal with). As an experimental physicist, such philosophical problems did not concern me directly. On the other hand, the post-war role of the Atom Bomb much worried me and it was Bohm's ideas about this problem which brought me and him together. Some physicists on the Project were narrowly fascinated by the idea of, after the war, constructing new nuclear bombs of much greater power. Bohm's imaginative concern enabled him to see the dangers of such developments. Like the great Quantum pioneer Niels Bohr, Bohm saw the need for holistic approaches with open dialogue between those with differing views. He saw that political analysis could not be avoided. However, after the war McCarthyist extremism grew and Bohm's recognition of the political aspects of nuclear policy led to his professional future in the United States being threatened. Bohm moved to Princeton where Einstein welcomed him. But Bohm was advised to leave the US because his passport might be taken away. He decided to leave the country.

After the war I worked on DNA which led me in 1952 to Brazil where I met Bohm again and was very glad to find him developing his ideas as before with creative courage and energy. A few years later he settled in England. He married an artist (as I did) and, in order to clarify ideas about a more unified culture, he began discussing with artists the relationship of art and science. During the 1960s environmental science began to make clear that science-based technology was creating serious environmental problems. This, together with the relentless background of the nuclear arms race, emphasised the need for a new way of seeing the world. Studying interconnections in our culture and developing an holistic philosophy was increasingly seen to be necessary and, in response, new cultural movements appeared. The British Society for Social Responsibility in Science (BSSRS) was formed and stressed the need for more sense of community in science and more integrated thinking about the problems of the world.

Following writing by novelist C. P. Snow on the gap between science and art, a BSSRS group was formed, including Bohm and myself, to examine the relation between art and science. We were stimulated by symmetry in architecture (e.g. by Buckminster Fuller) and in newly discovered viruses and giant molecules. BSSRS activities peaked in 1970; at a great conference which examined problems foreshadowing today's social and ethical concerns about molecular biology. There were many distinguished speakers, including Bohm on holism and dialogue.

In the US Charles Biederman, well known Structurist artist and author of considerable writings on the nature of art, became very interested in Bohm's ideas. Anthony Hill, a British artist with special interests in mathematics, had told Biederman of the importance of Bohm's new book *Causality and Chance in Modern Physics*. As a result in 1960 an enthusiastic correspondence began between Bohm and Biederman. On the other hand, in Britain, although Bohm's ideas attracted much attention in the 1970s and after, the general movement of idealism and libertarianism of the 1960s and 1970s faded. Political conservatism grew and monetarism increased its domination of the whole of culture. It was only after Bohm died in 1992 that I became aware of the remarkable Art/Science correspondence between Bohm and Biederman from 1960 to 1970. The correspondence illustrated clearly Bohm's ideas of dialogue and cast light on how Art/Science ideas could point the way to increased unity in all aspects of modern culture.

Today, with severe world-wide disunity in culture as a whole, it is useful to recall the surprising degree of Art/Science unity which has existed in the past. For example, the recent Tate Gallery exhibition "Turner and the Scientists" reveals how Turner painted breath-taking landscapes where beauty and spiritual force emerged with the aid of careful study based on science such as meteorology and geology. This was possible because the communities of science and art were then closely linked (as was symbolised by the Royal Society and the Royal Academy being next door to each other). Turner trained as a builder and architect, mixed with poets and scientists and admired steamships, railways and the Industrial Revolution. (See *Turner and the Scientists* by James Hamilton 1998.) The Romantic movement had by no means built a solid barrier between Science and Art. Similarly, seventeenth-century scientists had viewed science as a way of "Reading the Mind of God". And, leaving aside Science/Religion opposition, in the nineteenth century it was not unusual for astronomers to see the "Glory of God in the Splendour of the Heavens".

Today it is specially difficult to see science in a wide perspective. One reason for this difficulty is that research workers find themselves, like other professionals, under increasing pressure. But the need to see science from wider angles is greater than ever. The Bohm–Biederman correspondence gives unique illumination of Art/Science relations. It illustrates very well Bohm's ideas of creative dialogue and can help us move towards greater unity in the modern world. Such unity is of vital importance. The Bohm–Biederman correspondence demands most careful study.

M. H. F. Wilkins

PREFACE

Charles Biederman

It was sheer chance that I encountered David Bohm's writing in 1958, having ordered one of his books on speculation from a scientific catalog. I knew nothing about him. What struck me about his work and prompted my initial letter was his underlying effort to seek for some larger sense of reality, which seemed a very humanized search. This was further evidenced as our correspondence progressed, and he made the effort to communicate with me beyond the private language of physicists. I soon discovered that David was as concerned with the disparities within his own field as I was with those within mine. As a result, I remain indebted to him for the stimulus and the urgency he brought to our discussions of the differing reality pursuits of science and art and the humanities, although we arrived at our respective conclusions only after our correspondence ended. I am sure that David would have hoped that our mutual efforts would stimulate others to confront the larger, neglected problem since Leonardo da Vinci of our perception of reality as evidenced by thousands of years of art and then by four centuries of modern science. Now more than ever, it seems to me, the problem is inescapable, and its resolution is crucial to the well-being of both fields. Long ago Whitehead observed that Newton left nature "without meaning or value", and that the limitations of science were "perhaps too narrow for science itself". This, it seemed to me, was also David's view.

EDITOR'S INTRODUCTION

Paavo Pylkkänen

On March 6, 1960 the American artist Charles Biederman wrote to the physicist David Bohm in London. Biederman had recently read Bohm's book *Causality and Chance in Modern Physics* (1957) which had been recommended to him by the English artist Anthony Hill. For Biederman the book had been a "tremendous experience to read" and he wanted to let the author know this. What neither of them realized at the time was that this one-page, spontaneous letter was the first of over 4,000 pages of correspondence, to be written between March 1960 and April 1969! In this first volume the early part of this correspondence is published, namely the letters ranging from March 6, 1960 to April 24, 1962.

David Bohm (1917–92) was one of the foremost physicists of his generation and the author of many influential books. He is known in particular for developing an alternative interpretation for the quantum theory, the theory that deals with atomic phenomena. The first consistent interpretation of the quantum theory was proposed in the mid-1920s by Niels Bohr. It made the radical proposal that a deterministic description of the movement of individual quantum systems (e.g. electrons) was no longer possible, and even more strikingly, that we should give up thinking of, say, an electron as just a particle and rather see it as something that can exhibit both wave-like and particle-like properties, depending on the experimental situation. In all this a kind of *undivided wholeness* of quantum phenomena was emphasized. Bohm started off as a supporter of Bohr's point of view and even wrote one of the best textbooks of its time, *The Quantum Theory* (1951), along these lines. Gradually, however, Bohm grew dissatisfied with what he saw as the fragmentary features of the Bohr approach, and here he was influenced by discussions with Einstein in Princeton. He began to examine whether a fuller description of quantum processes was possible, and in 1952 published a consistent alternative hypothesis about the quantum domain. This hypothesis went strongly against the prevalent view by suggesting that individual quantum systems like electrons are accompanied and determined by a new type of field, that they have a well-defined existence as particles, and move deterministically along trajectories. Yet it made very explicit some of the most radical features of quantum theory, such as a holistic property known as non-locality.

The response to Bohm's work was divided. Given that Bohr's interpretation was taught as the only possibility – sometimes dogmatically – some were satisfied that Bohm had shown that an alternative exists. Others felt that Bohm's hypothesis was trying to avoid the philosophically radical implications of Bohr's view, such as indeterminism and a kind of ambiguity of quantum phenomena.

Bohm himself saw his 1952 "causal intepretation" as a starting point or a source of insight rather than as a final theory. But as many of his fellow physicists were against the theory and as he could not obtain new empirical evidence in its favour, he began to focus on different things, to some extent returning to Bohr's way of looking at the quantum theory. However, he started to reconsider the causal interpretation more seriously in the 1970s, partly as a result of the work he did together with his colleague Basil Hiley and their research students at Birkbeck College, London University. Bohm's last ideas on the interpretation are published in the book *The Undivided Universe* which he co-authored with Basil Hiley and which was almost completed before Bohm's death in 1992. Although Bohm's 1952 work was initially resisted, it is today, in various developed forms, increasingly seen as a source of insight into the fundamental problems of physics.

Since the early 1960s – which is the period when the correspondence with Biederman began – Bohm started to develop a general framework for physics which he later called the *implicate order*, known to the general public through Bohm's *Wholeness and the Implicate Order* (1980). One of the values of the correspondence is that it shows the development of Bohm's ideas leading towards the implicate order. Bohm also made significant contributions in many other fields, including philosophy, psychology and biology. In discussions with the Indian teacher J. Krishnamurti they explored human consciousness and the possibilities of its transformation. And in this correspondence, of course, we see him engaging in an intense dialogue about the meaning of art and artistic creativity. A biography that covers the various aspects of his work has recently been written by F. David Peat (1996), and a useful introduction to the more technical parts of his work can be found in the introduction to Bohm's *Festschrift* by Hiley and Peat (1987).

Charles Biederman (b. 1906) was the founder, in the late 1930s, of an art form first known as Constructionism and later, in the early 1950s, as Structurism. This art form, which is a kind of relief art organized geometrically, developed from painting but shares its three-dimensional properties with sculpture. According to the art historian and critic Donald Kuspit (1985), "the key to Biederman's works is the use of planar geometrical elements as irreducible sensory – color – units, and their integration into structures that seem dynamically evolving." Thus Biederman often describes his work as dealing with the "structural process level of reality". Biederman also wrote and theorized extensively on art, the most notable example being his monumental book *Art as the Evolution of Visual Knowledge* (1948) (referred to in the correspondence itself simply as the *Evolution* book). Although he never became part of the mainstream, Biederman strongly influenced some important English artists such as Victor Pasmore and Anthony Hill and also held some significant exhibitions, including a retrospective at the Hayward Gallery in London in 1969. Illuminating discussions of Biederman's works and ideas can be found in, for example, Kuspit (1977) and Craven (1980a, 1980b), as well as in various exhibition catalogues.

In the foreword to Biederman's Hayward exhibition catalogue, Robyn Denny (1969: 4) first notes that between 1936 and 1966, "Biederman carried through one of the most radical and single-minded attempts in modern art to formulate a new definition for art and its relation to the organic world" and then goes on to consider the reasons for Biederman's neglect. One suggestion is that many people know Biederman only through his theories which they do not accept, very few having seen his work. Another reason is that while "art in America focused on the idea of 'object' as an entity separated from the natural world, Biederman was systemizing the Structurist comprehension of nature as a source of art". Biederman's audience has been very specialized and connected by an ideology and "a belief in the evolutionary process in art derived from certain basic premises". According to

Jan van der Marck (quoted in Denny 1969: 5), the approach of this group is programmatic: "They are concerned with the technical and the conceptual issues in the making of their works. They have a mathematical or natural science orientation. Finally they have a social ethic that may strike the critic as utopic." This explains, of course, why Biederman was interested in books like Bohm's *Causality and Chance*.

Robyn Denny thinks that Biederman's *Art as the Evolution of Visual Knowledge* (1948) played an important role in Victor Pasmore's conversion to abstract art which, in turn "has always been regarded as one of the most glamorous landmarks in British postwar art" (Denny 1969: 5). For Pasmore and some other English artists Biederman's book was very relevant:

> Many artists have claimed Courbet and Cézanne, Monet and Mondrian as prime sources of their creative sensibility . . . But Biederman in so doing offered . . . a coherent, uncompromising and committed proposition for a radical alternative to painting and sculpture which received an affirmative response from this small group of English artists.
>
> (Denny 1969: 5)

Jan van der Marck notes that following Cézanne, Biederman is motivated by structural processes he observes in nature, especially outdoors:

> He studies nature's "building methods" as they present themselves in a dynamic interaction of form, space and light, and abstracts these insights into a non-mimetic art of pure, geometric form. By proposing the "structural process level" of reality as the source of art, Biederman offers an alternative to two types of art which he opposes equally: representational art, with its references to mimetic, visible reality, and non-geometric abstract art, as it favours emotion over reason.
>
> (van der Marck 1969: 12)

Here is how van der Marck describes a Structurist relief.

> [It] consists of a flat rectangular plane with an ingeniously organized array of smaller planes attached to both the background and to one another. Within the limitations of their geometric shape and parallel or perpendicular positions, these planes create a unique arrangement of volume and space.
>
> (van der Marck 1969: 12)

Robyn Denny notes that those who have not seen Biederman's reliefs before are often surprised at the intensity of colour and its optical complexity: "For Biederman, colour and light are as essential ingredients of each work as the substance and form of the structure itself." And if

Denny at times has found Biederman's arguments hard to connect with, his work appears free from any idiomatic constraints. Biederman's reliefs

> contain an extraordinary degree of rigour, discipline and control. A sense of order, a command of means, and uncluttered clarity of purpose. . . . There is no doubt in my mind that Charles Biederman has made a unique contribution to modern art.
>
> (Denny 1969: 6)

To honour Biederman's 90th birthday in 1996, the Tate Gallery exhibited one of his works which it has in its possession.

The above throws some light on the factors that are likely to have triggered the intense correspondence between Bohm and Biederman. Perhaps the most important was the interest that the two men shared in the natural world. The fact that they were approaching nature from different angles, those of art and science, meant that they complemented and in this sense needed each other in order to obtain a fuller understanding of nature. Another factor was their dissatisfaction with the dominant trends in their fields. This is evidenced by the fact that in his very first letter to Bohm, Biederman draws a parallel between the situation in art and in quantum physics. His idea was that in both art and in physics traditional views of nature had become inadequate in the late nineteenth and early twentieth centuries, and that in both fields the mainstream had failed to respond adequately to the new situation. For Biederman a favourite example of such a failed response was Surrealism and for Bohm it was the "Copenhagen" interpretation of the quantum theory. A further common factor was that both Bohm and Biederman were proposing new and what they felt to be more adequate views of nature in their respective fields. Finally, their opposition to the mainstream and persistence in developing and holding alternative views did not help them to gain respect amongst their colleagues – the label of the maverick was appropriate to both men.

When the correspondence began, both men were already mature thinkers and practitioners in their fields and consequently the knowledge they draw upon in their letters is extensive. This is not, however, to say that the letters are inaccessible to the general reader who may not possess a detailed knowledge of Bohm's and Biederman's work. On the contrary, the fact that we have here a physicist writing to an artist and vice versa means that difficult ideas are often explained in a clear and simple fashion. It is this that makes the Bohm–Biederman letters much more accessible and relevant to the general reader than a correspondence between theoreticians usually is. It will, of course, be helpful to have some familiarity with their work prior to the correspondence. Thus, for example, the reader may find it useful to have Bohm's *Causality and Chance in Modern Physics* to hand when reading the letters – this was, after all, the text that prompted Biederman to write to Bohm in the first place. Biederman's writings are harder to obtain, but the articles mentioned in the bibliography at the end of the book provide a fuller introduction to his work which the general reader is likely to find helpful.

CREATIVE DETERMINATION

Bohm and Biederman discuss a wide range of topics in the letters that follow. The fact that the form of the discussion is that of correspondence means that there is a great deal of spontaneity in the way the different topics come up. This makes it more difficult to summarize the discussion and to find the main topics and themes. Fortunately, Bohm himself makes a summary of this kind, notably in his December 22, 1961 letter (L16, DB8), and to a lesser extent in his April 15, 1962 letter (L30, DB16). The reader will find these summaries useful, although they are, of course, written to Biederman rather than to someone with no previous knowledge of the correspondence.

Certainly one central topic in the correspondence is the relationship between human beings and nature, in particular as this has to do with artistic creativity and scientific endeavour. I also think one is justified in making the following simplified contrast between the overall styles of the two thinkers: as a scientist Bohm is more concerned with understanding nature as it exists independently of human beings, whereas Biederman as an artist is more concerned about the meeting of human beings and nature, and properties that come into being as a result of that meeting. Through his scientific and philosophical thinking, Bohm aims first to formulate a vision of nature as a creative process and then to understand human beings and human creativity in terms of that vision. Biederman's Structurist art, in turn, focuses on the special role of human beings: the Structurist artist discovers and reveals those creations inherent in nature's process which only human nature can actualize.

This basic initial difference of approach shows itself in many of the individual debates. For Biederman, nature considered without human beings is determinate and there is no room for properties such as chance, indeterminism, disorder, randomness, lawlessness, etc. as objective. These only arise in the meeting of human beings and nature and they are projected on to nature by us. This explains why Biederman approves of Bohm's argument in *Causality and Chance in Modern Physics* in which he opens up the possibility that the quantum level, currently the basic level of nature, can be seen as deterministic, contrary to what the prevailing "Copenhagen" interpretation of the quantum theory had claimed. It appears that Biederman prefers the view of nature – when understood independently of human beings – as relatively simple, determinate and straightforward and sees complexity as arising only when human beings and nature meet.

Those who view Bohm as a simplistic determinist – and there may still be quite a few – would expect him to have approved of Biederman's attitude. Bohm's overall response to Biederman – as well as much of Bohm's published work since the mid-1950s at least – reveals, however, that his concept of "human-independent" nature is far more complex and permeated by features such as indeterminism and contingency than that of, say, Biederman. This state of affairs is indeed captured by Biederman's remark to Bohm: "What you want to attribute to nature, I attribute exclusively to man" (L11, CB6, December 28, 1960). In a number of letters we find Bohm trying to convince Biederman how various properties such as chance, randomness, contingency, possibility, etc. can be seen as objective properties of human-independent nature. This ought to surprise those who have seen Bohm as a simplistic determinist, perhaps based on their reading of Bohm's 1952 papers on "hidden variables". For here we have Bohm, who is internationally known as a defender of a deterministic interpretation of the quantum theory, and thus for many a defender of strict determinism in nature, arguing strongly for the objective existence of properties such as contingency, chance, indeterminism, etc. Of course, Bohm does this already in *Causality and Chance*, but here the point is made more vividly, given that Bohm is defending the role of indeterminism rather than questioning it, as he most famously did in his 1952 papers.

Regardless of the differences in their approaches, we find the two thinkers developing many common themes and ideas as well. An important idea is that of *creative determination*, which Bohm and Biederman see as different from and going radically beyond either mechanistic determinism or pure chance and chaos. A related and very interesting discussion has to do with the concept of time. Bohm summarizes their discussion as follows in his December 22, 1961 letter to Biederman (L16, DB8):

> Each moment traces its past and projects its future in a unique way. At each moment there is an infinity of possibilities, which (we both agree) have, in some sense, a real existence. You argue that man can choose from these possibilities, and nature by itself cannot.
>
> The past is not by itself fully determinate, but is ambiguous as to what it is (or was). The future helps to determine what the past really was. Thus, the past does not fully determine the future. (If only because in some measure, the future partly determines the past.)

It is through arguments such as the above that Bohm and Biederman try to make room for a new type of "creative determination". One idea is that if the past does not fully determine the future, this leaves room for the present moment to make a creative contribution. This discussion also relates to the perennial philosophical debate between determinism, freedom and responsibility. If the laws of physics completely determine the behaviour of our bodies, how can we be held responsible for our physical actions? And even if we allow chance and indeterminism in the laws of physics, what difference would it make? For then the behaviour of our bodies could be governed by, say, a combination of deterministic and random laws, and this does not seem to capture what is meant by responsible action either. In trying to develop a concept that goes beyond either pure chance or strict determinism, Bohm and Biederman are thus tackling not only the question of artistic creativity or creativity in nature, but also the problem of understanding how responsible action is possible if we are physical beings.

In my view their discussion of creative determination and time brings up fresh viewpoints that are often neglected by philosophers. For example, when philosophers discuss issues such as determinism, they often bring relatively simplistic notions of physics into the discussion. Bohm's suggestion that the past is ambiguous is related to the uncertainty principle of the quantum theory. The idea that the present moment can make a creative (rather than random or strictly determined) contribution has thus to be understood within the wider, quantum-theoretical idea of the "ambiguity of the mode of being of parts" which Bohm explains in his second letter to Biederman. This is not to claim that Bohm and Biederman have come up with definite solutions of important problems in their letters. But they certainly discuss some very difficult issues in novel ways, which may well benefit contemporary researchers struggling with these topics.

To help the reader deal with the vast amount of information in the correspondence, at the end of the book I have provided a brief summary of each letter. These summaries are inevitably selective and thus by no means do full justice to the various letters. They should be taken as a rough guide for orienting around the material rather than as a precise summary and evaluation of the contents of the letters.

Personally I think that the Bohm–Biederman letters are an extraordinary treasure. We live in an age where science and technology increasingly need to serve the interests of economic growth and immediate practical applications. The adverse effects of science and technology are well known, but it seems to be impossible to control and foresee developments such as those in information technology. For example, how will the current developments in this technology affect human consciousness? Will consciousness have to mechanize itself in order for the new information technology to function efficiently? Similarly, it is hard to see the role of art in the "postmodern" era. Much of "serious" contemporary art is inevitably linked with commercialization and mass-media culture. The Bohm–Biederman letters provide a space in which one can critically reflect these contemporary developments and glimpse alternative directions.

REFERENCES

Craven, D. (1980a) "The Art of Charles Biederman", unpublished PhD thesis, University of North Carolina, Chapel Hill.

Craven, D. (1980b) "Charles Biederman's Art of Complicity", *Arts Magazine*, March, pp. 130–5.

Denny, R. "Introduction", in *Charles Biederman: A Retrospective Exhibition with Especial Emphasis on the Structurist Works of 1936–69* (exhibition catalogue for the Hayward Gallery, London, and the Museum and Art Gallery, Leicester), Arts Council of Great Britain, London.

Hiley, B. J. and Peat, F. D. (1987) *Quantum Implications: Essays in Honour of David Bohm*, Routledge, London.

Kuspit, D. B. (1977) "Charles Biederman's Abstract Analogues for Nature", *Art in America*, May/June, pp. 80–3.

Peat, F. D. (1996) *Infinite Potential: The Life and Times of David Bohm*, Addison Wesley Publishing Co. Inc., Reading, Mass., and Harlow, England.

Van der Marck (1969) "Biederman and the Structurist Direction in Art", in *Charles Biederman: A Retrospective Exhibition with Especial Emphasis on the Structurist Works of 1936–69* (exhibition catalogue for the Hayward Gallery, London, and the Museum and Art Gallery, Leicester), Arts Council of Great Britain, London.

(*Note*: For a full listing of works by and on Bohm and Biederman, please refer to the Bibliography at the end of the book.)

I

(1)

March 26, 1960

Dear Mr. B.

Thank you very much for your letter of March 6, in which you express a good opinion of my book, CAUSALITY AND CHANCE. I am particularly happy that you, as an artist, found the ideas in the book interesting. I too have been concerned with the gap between science and other fields of human endevour. I regret this gap as bad both for the scientist and for the non-scientist. On the one hand, the general conceptions of the scientist are strongly influenced by general ideas that are, so to speak, "in the air" in any given period; and on the other hand, science contributes ideas which strongly influence the thinking of subsequent periods.
There is no doubt that the currents of thought, emotion, and general reaction to social and world conditions, which led to positivism and existentialism in philosophy, to surrealism, expressionism and action painting in art, to similar trends in literature, etc., have contributed heavily to the ideas behind the Bohr interpretation of the quantum theory. Nevertheless, the problem is very complicated, because all these reactions contain aspects of truth, inexplicably mixed up with a great deal of falsehood. For example, in science, there really has been a breakdown of older ideas on causality, space-time, etc., and Bohr's point of view expresses certain aspects of this breakdown correctly. Similarily, existentalism exposes certain weaknesses and hypocricies that are really in our social and personal relations; and it raises in a sharp form the problem of what can be meant by freedom. In art, the older classical forms no longer express modern reality, and some new development seems to be needed. Similarily, in literature, the form of the novel does not seem to be adequate to expressing the complicated and ambiguous character of the individual today, which results from his new social relations.

The pity is that the kernel of truth in all of these movements has swept many people along into swallowing a tremendous number of false aspects at the same time. (One could say that people find it difficult to keep the baby without also keeping the bath water.) Thus people are led to adopt inadequate, excessively negative attitudes to themselves, to their fellow men, to society as a whole, etc. In this way they contribute to worsening the very breakdown that they are reflecting.

What is missing is a new over-all point of view that is adequate to our new situation, our new relationships, and our new knowledge. We cannot ever return to the old ideas, but it is important also not to throw up our hands in despair and to adopt irrational points of view.

I am looking forward very much to reading your article when it arrives.

Very sincerely yours,
David Bohm

A NEW VISION OF TOTALITY

Red Wing
Route 2
Minnesota, USA
March 6, 1960

Dear Professor Bohm:

I believe that whenever one comes across a book that has become a tremendous experience to read, one should let the author know it. This is what your book, *Causality and Chance*, has been for me. I read it just recently, and I will have to read it again.

To explain my interest in your book. To put it briefly, the notion of indeterminism has always seemed contrary to experience, which, even after reading your very fine book, I cannot accept even as an eventually limiting case.

It seemed an almost inescapable error that the early mechanistic outlook should have arrived at an absolute. But I think there is something else afoot when the complementary-indeterministic orientation also arrived at an absolute. For the latter takes place in a period when there is a general awareness of the non-absolute character of our structural knowledge about nature. When the first mechanistic outlook fell, there had been no crisis in science but just the opposite; afterwards there was the tremendous effort to re-orient. In the present stage of the mechanistic view, there came a genuine crisis, and it appears to me that complementary-indeterminism is but a rationalization of that crisis.

This effort impresses me as the Surrealism of science. When the old view of nature was no longer adequate for the further continuance of art, the Surrealists, unable to wrest a new order from nature, assumed the view of nature as "disorder." I sympathize with your belief that a deeper penetration will reveal a nature of causality. But there is the possibility that this will also dispel the basis for the present "lawless" view of nature and, rather than make it a limited case, will dispense with it entirely.

But I must tell you why I feel so deeply the reading of your book. It is because in general your view of nature appears to substantiate the direction in which I have been considering these problems, not from the view of a physicist, but from that of an artist. I have been concerned with the hiatus that exists between art and science, the disparity and separation that exists between these two fields. And, what better place to get at the problem than to consider their respective orientations to nature as reality. The view I have gradually been approaching is one that appears to be enforced by some of your conclusions.

Three years ago I wrote a long essay on "Mondrian and Science." It seems very likely that it will finally reach publication sometime this year. If you will permit me, I would like to send you a copy when it appears.

May I thank you for making your views possible for me to read. My feelings for hope for the future are increased by such as you.

Sincerely yours,
Charles Biederman

3, Berkshire Road
Bristol 7
Glos., England
March 26, 1960

Dear Mr B.:

Thank you very much for your letter of March 6, in which you express a good opinion of my book, *Causality and Chance.* I am particularly happy that you, as an artist, found the ideas in the book interesting. I too have been concerned with the gap between science and other fields of human endeavour. I regard this gap as bad both for the scientist and for the non-scientist. On the one hand, the general conceptions of the scientist are strongly influenced by general ideas that are, so to speak, "in the air", in any given period; and on the other hand, science contributes ideas which strongly influence the thinking of subsequent periods.

There is no doubt that the currents of thought, emotion, and general reaction to social and world conditions, which led to Positivism and Existentialism in philosophy, to Surrealism, Expressionism and Action Painting in art, to similar trends in literature, etc., have contributed heavily to the ideas behind the Bohr interpretation of the quantum theory. Nevertheless, the problem is very complicated, because all these reactions contain aspects of truth, inexplicably mixed up with a great deal of falsehood. For example, in science, there really has been a breakdown of older ideas on causality, space–time, etc., and Bohr's point of view expresses certain aspects of this breakdown correctly. Similarly, Existentialism exposes certain weaknesses and hypocrisies that are really in our social and personal relations; and it raises in a sharp form the problem of what can be meant by freedom. In art, the older classical forms no longer express modern reality, and some new development seems to be needed. Similarly, in literature, the form of the novel does not seem to be adequate to expressing the complicated and ambiguous character of the individual today, which results from his new social relations.

The pity is that the kernel of truth in all of these movements has swept many people along into swallowing a tremendous number of false aspects at the same time. (One could say that people find it difficult to keep the baby without also keeping the bath water.)

Thus people are led to adopt inadequate, excessively negative attitudes to themselves, to their fellow men, to society as a whole, etc. In this way they contribute to worsening the very breakdown that they are reflecting.

What is missing is a new overall point of view that is adequate to our new situation, our new relationships, and our new knowledge. We cannot ever return to the old ideas, but it is important also not to throw up our hands in despair and to adopt irrational points of view.

I am looking forward very much to reading your article when it arrives.

Very sincerely yours,
David Bohm

Red Wing
Route 2
Minnesota, USA
April 11, 1960

Dear Professor Bohm:

Thank you very much for yours of March 26. I read your letter with considerable interest, and I regret that it is my misfortune that I cannot talk with you on the problems that are of mutual interest.

You were probably being kind to me, but I had hoped you might respond directly to some of the statements I made dealing with the subject of your book. If I am saying what seem to you foolish things about science, I'd feel you were doing me a service to say so. Unfortunately I do not know any scientist with whom I can discuss problems of science and art, let alone one of your competence. On my own I read what science books I can afford, and hope each time that I will be able to comprehend something of their contents. This reminds me to tell you that I thoroughly enjoyed the clear way in which you wrote your book, the lack of pomposity, and the very genuine devotion to your work reflected in your words. Except for the mathematics I think I will be able to comprehend most of what you write, by reading it until I do.

There is one thing that has always remained a mystery to me about the use of all such terms in science, as random, lawless, indeterministic, disorder, etc., etc. How does one distinguish disorder from order in the structure of nature? It appears to me that when the physicist describes the variable activities of the individual atoms to all the others observed, he does no more than produce a *description*. He then proceeds, as I see it, to take what remains a descriptive abstraction and offers it in terms of an inferential abstraction. If this is the case, one is not being given any empirical evidence, but only the manipulation of words where there is a confusion of descriptive with inferential abstractions.

I was extremely interested in the subject of your letter, and the wide view you have of it. Would you be interested in writing an article, expanding these notions for an art magazine? A friend of mine in Canada is starting a new magazine, it will be called *The Structurist* and devoted to that contemporary art that followed out of the Dutch school of De Stijl. We are anxious to have articles by scientists such as you who can write along the lines you take in your letter to me. So I hope you will be interested. Have you published anything along these lines?

I was particularly struck by your statement that Bohr's formulations were greatly influenced by what was going on in other fields of activity than his own. Could you be more specific? For this seems like something of extremely vital importance which should be publicly made available. In my article mentioned in my previous letter, I am mostly concerned with the adverse effects science has had on art, because of the narrow concern of scientists and the erroneous use of their theories by artists. I also state that art has failed to exert its beneficial influence upon the course of science, the opposite of that aspect you mention in your letter.

You write that "some new development seems to be needed" in art. Since 1937 I have been working and gradually formulating such a development. In 1948 I published the results of ten years of writing, to indicate by means of evolving history just what sort of new development was potential to our times. I am sending you a copy of this book. In it you will discover that I cannot agree with you, as regards Surrealism, Expressionism and Action Painting, as possessing what you call a "kernel of truth." To my view, these and most other art movements of our century, like political and other social forces, have used certain kernels of truth as the means for escaping the extremely difficult problems of our times, and resorting to pathology in art and barbarism in politics, solutions of violence whether with a brush or guns.

Since my 1948 book appeared, a number of artists have gone in the general direction I formulated, especially since about 1950. These artists are English, Dutch, Canadians and Americans. I am enclosing three issues of their magazine published in Amsterdam. All these artists are concerned with the relation between art and science, as you will see they represent differing viewpoints on both subjects. I would welcome your severest criticism of my essay "Art and Science as Creation" appearing in one of these issues. I am also sending you a copy of my most recent book, *The New Cézanne*, which, in effect, has the purpose of showing the transition from the old mimetic view of nature to the new view of nature as a creative process.

My best wishes,
Charles Biederman

3, Berkshire Road
Bristol 7
Glos., England
April 24, 1960

Dear Mr Biederman:

Thank you very much for your letter of April 11. I am awaiting the receipt of your book and articles with great interest. I am afraid however, that I cannot write the article you suggest for *The Structurist*, mainly because I really know very little about art and do not feel qualified to write on the subject, or even on the relation between physics and art. Perhaps after I read your book and articles we might discuss this question further.

As for Bohr's being influenced by what was going on in other fields, he admits (privately) that he was strongly affected by Kierkegaard's Existentialism. The point of contact is in the question of the dual nature of subject and object, of action and contemplation, etc. In physics, there arises the analogous problem that the observing apparatus (which is the proxy of the observing subject) is linked to the rest of the universe by indivisible quanta. Indeed, this is only a special case of the universal linkage of each part of the universe to the whole in the same way. Such a linkage means that the essential sides of the very being of each system are in its indivisible quantum relationships with what it is not (i.e. what is other to it). This holds as much for the observing subject (or for his proxy, the observing apparatus), as it does for any object that we might care to distinguish from its background. As a result, there is an inherent ambiguity in what each partial system *is*, because what it is is in essential respects, its relation to its other. In this way, physics has led to the denial of the mechanistic idea of a universe made up of distinct parts (or elements), such that each part exists separately and simply comes into interaction with other parts. This idea of ambiguity of the mode of being of parts is, in my opinion, a very deep one, and of fundamental importance for further progress. It is very significant that even in physics, the science where analysis into parts is developed to the utmost, one discovers that this is, in fact, not a correct description of reality.

The connection with Existentialism is fairly clear. The Existentialists are concerned with the question of being vs. existence. They, in my opinion correctly, feel that there is a wide gap between our concrete existence and our apprehension of this through ideas which describe what we *are*. When we look deeply into ourselves, we find that all our usual well defined ideas about ourselves become arbitrary, unnecessary, fade into nothingness (much as happens in the atomic domain when we try to follow an atomic process in great detail). People who have reflected long (perhaps too long) on these questions, quite justifiably begin to feel anxiety, fear, nausea, etc., because the whole foundation of their being seems to lose its solidity and its value. We discover that the question of what we *are* is, in considerable measure, ambiguous. And it is not surprising when we think of why this should happen. For in such introspection, one part of the self separates itself and tries to look at another part. In a rough analysis, it can see certain outlines of characteristics, more or less. But as a closer inspection proceeds, ambiguity inevitably appears, because an essential side of the self that is being watched is in the self that is watching, and of course, vice versa. Thus, it is just as wrong to split the self into observer and "objective" self as it is to split the world in two. This is a split that is OK for some restricted purposes, but it is wrong as an expression of the universal and necessary laws of being of the universe or any part of it.

Thus far, then, I find myself in agreement with Bohr. Where I disagree is in his contention that we can merely accept this state of affairs as an irreducible fact, and that we cannot make any further progress towards comprehending how everything (including ourselves) exists in this indivisible relation in which its full being is only in the whole. I think that it is possible to go further and to develop a mathematical and physical theory which gives us some conception of this new relation of part to whole. Perhaps I could say we need a kind of "vision of totality", mathematical and perhaps ultimately, artistic, literary, social, etc. as well. This vision would (like all our partial conceptions) distort and leave out a great deal. Nevertheless, it might carry us a step further in our understanding of the problem.

This new point of view depends on understanding the world and each part as infinite. Here, it is not merely quantitatively infinite, for such an infinity is still limited to repetition endlessly of a certain quality. Nor is it merely qualitatively infinite. For this is still limited to the need for endless change of quality. True infinity cannot be limited by anything else at all. Either we can say that it has no "other", or else that the infinite is its own other. This amounts to saying that true infinity is self-limiting, or that it is free (in the sense that there is nothing outside it to limit it).

Here, it is important to stress that if the infinite did not limit itself, it would have no being at all. For whenever we have to do with something, there must be a limit (e.g., it is not something else). Hence, it is essential to define the infinite as self-limiting, and not just as the unlimited, which would be nothing at all.

This is where the finite comes in. The finite is the limit *in* the infinite. (It is important to avoid saying "limit *of* the infinite" as this would deny that the infinite is really infinite, i.e., unlimited by anything outside of it.) The illusion that is common to most thinking is that the finite exists in itself (i.e., that a particle exists in a certain part of space). Actually the finite exists only as a side or aspect of the infinite totality. We get into serious errors by imagining that certain entities (e.g., electrons or men) are *only* finite. In reality, everything is basically infinite and only secondarily, finite.

Here, we have an interesting new feature which first appears in the mathematics of the infinite. Thus, the number of points in any region of space is infinite, no matter what size this region may be. It is therefore possible to map the whole of infinite space into any part of it. In other words, to each point in the larger space corresponds a point in the smaller one. *Hence, every region of space, however small, can reflect the whole.* Even more, the relation between each part and the whole can also be reflected in the part. Thus, even the simple quantitative infinity of space has the new feature that *not only is each part in the whole but so also can the whole be in each part* (in the sense of a reflection, at least). This feature is reminiscent in my opinion, of certain features of some paintings, where each part reflects other parts and even the whole, with regard to colour, form, composition and other elements which go to make up the picture.

The above idea of the infinite shows already the breakdown of the notion that the world can be divided into separately existing parts. For already, even in this very simple view, an essential aspect of what each part *is* is that it reflects the other parts.

The further extension of the idea of infinity to time leads to an even more radical change. Here, what is suggested is to reverse the usual idea of first imagining time and then saying that things exist and move in time. Rather, we begin with existence and process, and say that time is the order in this process. Thus, we define each time concretely as the "time when" such and such existed, or changed, and each position as the "place where" it was, etc. To carry through such a view consistently, we should begin with the concept of totality, which is infinite and eternal. This includes all that there is, was and will be. If we knew this, we would know all reality. We would know every concrete existent, every law (relationship) and the limits of every law. Of course, we can only select out certain aspects of this totality. Each science reflects some aspect, the arts another, the poet another, and so on. Even more fundamental is the split corresponding to past and future. At any one moment, one side of eternity is reflected as "past", i.e., it is gone. Another side is the future, which is yet to come. On the basis of some limited knowledge of the past, we try to project our knowledge, with limited success, towards the future. Actually, what is present now is neither the past nor the future, but a reflection of these, which if interpreted properly would tell us, in principle, all about them (only that this would require an infinite effort of interpretation). The split that we make at each moment is to divide all existence, at that moment, into two sides, one reflecting that which has been, the other implying that which will be. The reflection of that which has been is what is available to us immediately. Nevertheless, both sides actually exist together in each moment, they combine to make a totality. As a result, one side implies the other. I suggest here an analogy to advertising signs made of flashing coloured electric lights. In this way pictures are made, let us say with red and blue lights. Suppose we could only see the red lights (which are analogous to the reflection of the past). Then the general outline of the blue parts of the picture would be implied by the red parts, but its details would be missing. Thus, we would know something of the "shape of things to come" but not very much in detail.

I suggest that the above is the way in which we apprehend selected aspects of reality. They are selected by our location in space and time, by the nature of our senses and interests, by the character of our instruments, knowledge, skills, techniques, etc. Everything that we know is a selection out of an infinite totality. But here we must be careful. *Even our knowledge as well as our thoughts, feelings, etc., are parts or sides of this totality.* Remember here the point made by Niels Bohr, that anything less than the totality is, in some extent, ambiguous in its mode of being. Thus, even if we consider the whole universe, but leave ourselves out of it, we will leave essential ambiguities in the picture. For just as much as the mode of being of each man is completed only in his relationships to other people and to the rest of the world, so also is the mode of being of other people and the rest of the world complete only in their relationships to that man. This shows up, for example, in that nature without man is very different from nature with man in it (and transformed by man). Each man may contribute something essential to the being of others, and even to nature. Of course, usually it does not seem to happen this way, but this may be because we do not understand ourselves. Especially, we do not understand that each man is potentially infinite, and that if he were able to realize himself as such, he could have an enormous effect on other people, simply by being what he is (because it could become clear to other people in this way that they too are potentially infinite).

The above point of view implies that the notion that there exist "things" that are *only* finite is at best, a crude simplification, good enough for some purposes, but basically wrong in its overall implications. Since the very idea of a "thing" implies limitation and finiteness, we could say that in a sense, there is really no "thing" in the world. Even chairs, tables, etc., which seem at first sight to be limited "things" are not really so. If we apply a little heat, the molecules go into movement and they dissolve away into gas. Put them into an atomic pile and even their atoms would dissolve away into other atoms and energy. At very high energies all kinds of particles transform into each other. Thus, there are really no final or eternal limits on what any "thing" actually is, or can become. Such limits are only of relative validity, i.e. *the limitations are themselves limited in their domains of validity*. Thus, even the simplest "things" are really infinite, and much more so, the same holds for man, who is infinitely more complex and subtle in his make-up.

Considering the above, it is therefore not surprising that man's first impression of infinity is a sense of "nothingness". Thus, the empty night skies suggest infinity. What this really means is that all the apparently solid and everlasting features of nature, society, ourselves, etc., are realized to be founded on essentially nothing at all (or at least what is essentially nothing in relation to the totality). Every one of these things has something in it arbitrary, contingent, and therefore it is subject to being altered or destroyed by what is outside of it (as well as perhaps by its own internal laws of development). Even what seemed to us the most eternal values are seen to partake of this nature. It is therefore not surprising that this view has much in it that is frightening to all of us.

The other side of the picture is, however, that infinity denies the self-existent reality only of the finite. It does not mean a bare and empty nothingness at all. Rather, it means that no finite form is everlasting or complete in itself. This can be understood as the certainty that we must eventually lose everything that we have, or are. But it can also be understood as infinite freedom. We are not bound *permanently* by anything in the past (even though we must take the consequences of the past into account in all our calculations). There is a real freedom in the universe, in the sense that every form of necessity has its limits. If we know those limits, we can take the appropriate actions and remove ourselves from the dominion of any particular kind of necessity, thus becoming free, at least in that respect.

Engels has said that "freedom is the recognition of necessity". On the basis of the above discussion, I would argue that this is a misleading simplification, because mere recognition of necessity does not liberate us. For example, we all recognize the necessity that we must die, but we go on dying whether we like it or not. To be free of death we would have to understand the limitations of the laws that make death necessary, and to take the appropriate steps to remove ourselves from the dominion of these laws. We are at present able to do this in a small way. Thus, some of the causes of death are the actions of bacteria. If we take steps to remove ourselves from subjection to bacteria, we free ourselves from some of those things that make death necessary. But the more fundamental origin of the necessity of death is in our own internal processes that constitute our mode of life. And these are what we must understand if we are really to prolong life in a fundamental way.

You may now ask what is the positive content of freedom? In other words, after we have removed ourselves from the dominion of necessity, what will we do? The answer to this question is I think, closely related to the nature of art. We are really asking ourselves, "What *are* we, most essentially, after we have removed all external factors that now limit us?" I suggest that each one of us is something infinite, which at least reflects the infinite totality. This something should itself be a kind of whole. That is, it should have the kind of completeness, unity, integral character, etc., that is in a good picture. In such a picture, one cannot analyse the whole into separately existing parts (e.g. spots of paint etc.). Rather, to the extent that the picture is a real work of art, the justification of each part is only in the whole picture. Similarly, with regard to our own existence in time and space, we do not say that the past completely *determines* the future, for this would deny novelty to the future and wholeness to our existence in time. Rather, we say that the past in some way *limits* the future (remember the analogy of the picture made of flashing lights, in which the past shows "the shape of things to come" but not its details). Nevertheless, there is room for something new to exist in its own right. The freedom of each new thing to be what it really is, is limited by the past. This limitation applies to ourselves as well as to everything else in the world. But insofar as we understand these limitations, we can remove ourselves from their dominion and we can, in a sense, make room to allow us to be what we are, i.e. a complete, whole person, not only at one moment, but over a whole lifetime. In such a mode of existence there is genuine freedom, in the sense that the future is not fully determined by the past any more than, for example, the left-hand side of a picture is fully determined by its right-hand side. Rather, both sides simply make up the whole picture, in the sense that, while they limit each other in certain ways, this limitation is incomplete when the whole picture is not taken into account.

Of course, in doing all this, we cannot ignore causality, any more than the painter can ignore the requirements of technique, the properties of paint, etc. It is just by understanding causality very deeply that we can find its limitations, which leave room for us to be free, as the painter by mastering his materials and techniques finds room to express freely something that is an essential side of himself. By the word "express" one does not mean that this something is already there in him, fully formed and just waiting to come out. Rather, one means that it comes into this form fully only in the process of expression. Thus, the process has in it an aspect that is free, not determined by anything other than itself, i.e. it has the character of the infinite to be self-limiting (or perhaps, to be "creative", it could be said).

All of this raises another interesting point. A good picture is not only an integral whole, but even more, it achieves this wholeness by expressing something having universal significance. In other words, while it is something specific, particular, limited in its existence, etc., its relationships to its parts are rich enough and of the right character to suggest the universe and its relationships to its parts. Thus, it is "universal" in the sense that somehow its structure reflects that of the universe. In other words, it makes a kind of "world in itself". But besides this, it must make clear the other side of the truth, viz. that the picture is also not a world in itself, but obtains its full significance by its relationship to what is outside of it. In this way, it obtains another kind of universality. And I think that each truly existent thing does the same (a human life that could be lived freely, for example). In its structure, there is a wholeness that suggests a world in itself, yet this wholeness is not self-existent, but rather exists only in its relationship to the complete whole.

I am afraid that I have been getting a bit out of my field in the above speculations. Nevertheless, I would appreciate hearing how they strike you as an artist.

I have gone a lot further toward developing these ideas in mathematical form, but they are still only in a preliminary stage of formulation. Perhaps we could discuss them further after I hear from you about what I have written so far.

Perhaps it would be helpful if I summarized the main idea I have been developing so far. It is based on the notion that anything less than the infinite and eternal totality of all that there is, was or will be, is inherently ambiguous as to what it is, because essential aspects of its being reside in its relationship to this totality. This ambiguity applies in all divisions that we make (subject–object, universal–particular, one science vs. another, science vs. art, etc., etc.). The most fundamental division is time itself, which divides this totality (at each moment) into two parts, that which has (at that moment) passed away and that which is (at that moment) yet to come. Each new moment constitutes a new division of this totality, containing a reflection of all previous divisions.

At first sight, we tend to conclude that the past and the future are each well defined in their being, with no ambiguity as to what they are. But a more careful analysis shows that essential aspects of the being of each are in their relation to the other. Thus, there is some ambiguity in past and future. We experience this ambiguity in certain ways directly. For when we try to say "now", we find that by the time we have said it, the time that we meant is already past, and no longer "now". And if we try to do it with clocks, so as to be more precise, quantum theory implies that a similar ambiguity would arise because of the quantal structure of matter. In fact, there is no known way to make an unambiguous distinction between past and future.

If we follow through the consequences of this ambiguity, we see that there is room for genuine freedom. For it becomes impossible that the past shall completely determine the future, if only because there is no way to say unambiguously what the past really *was* until we know its future. In other words, as in a work of art, each part acquires its full meaning only in its relation to the whole. To a certain extent, we are led to a conception of being, which cannot be specified unambiguously at one moment of time.

My disagreement with Bohr is that he stops with just giving the limitations on the older mechanistic and deterministic points of view, and discourages the development of new categories of thought concerning space, time, existence, etc., in which one will see the reasons for these limitations. Thus, if determinism is limited, this need not be a complete mystery. It may only mean that we must have a different idea of what things *are*, which shows quite naturally why determinism is limited. If we get a different idea of time and of what is to be meant by existence in time, then we see a natural reason why the past would not determine the future (e.g., as has been suggested, the law is not that the past shall determine the future completely, but rather, that the whole shall form a unity, as in a picture or a musical composition).

As you may perhaps have noticed, my ideas on determinism and indeterminism have developed since I wrote *Causality and Chance*, although what I now think about these questions was to a considerable extent, implicit in the point of view expressed in the book. To relate this to what you said in your first letter (March 6), I would say that neither determinism nor indeterminism (causality or chance) is absolute. Rather, each is just the opposite side of the whole picture. Wherever there is one of these categories, there must also be the other. Our method should be to begin with something that goes beyond both of these categories, viz., the infinite and eternal totality. We want to get to know what the totality *is* and how indeterministic it *is*. The role of indeterminism is merely to describe the fact that causal relation in time does not exhaust the whole of being. It does not mean absolute lawlessness, but only that any particular chain of lawful relationship is limited, i.e. not completely universal in its domain of validity. These limits leave room for new relations and new kinds of totalities to come into existence. In terms of a sufficiently broad context, all laws and all limits to these laws are seen to follow from the fact that the whole (which includes time as well as space) is indeed a kind of unity.

With regard to your letter of April 11, you ask about the use of terms in science such as "random", "lawless", "indeterministic", "disorder" etc., "How does one distinguish order from disorder?" You are quite right in supposing that there is a great deal of arbitrariness and confusion here. Nevertheless, I believe that underneath it all, there is a real problem that remains to be solved, and that it is not just a question of "descriptive abstraction". For example, in the question of determinism vs. indeterminism, there is as I have said, a necessary complementary relation of the two ideas. Each event can always be studied (a) as unique and singular, at least in some aspects, and (b) as part of a causal chain. To be unique and singular, there must be something in it which is not just a reflection of other elements in the chain, e.g. its past, although it may, in part, be such a reflection. Similarly, every chain of determination has a beginning and an end. From the point of view in which we consider this chain, we must say that here this particular form of determinism becomes irrelevant, and we have a lack of such determinism. Likewise we may consider elements which are, in a narrow context, unique, singular and not essentially related to each other. In a broader context (e.g. space and time) one will see them coming into relation, as necessary parts of the broader whole. Thus, each form of indeterminism is also limited. Similar ideas can be developed with regard to order vs. disorder, randomness vs. law, etc.

With regard to your view concerning Surrealism, Expressionism and Action Painting, I agree in considerable measure with you about them. I personally do not like them, nor do I regard them as really healthy trends in art. However, where I differ from you is in my evaluation of their significance. I think that many of their proponents have honestly been trying to solve real problems. Of course, they have attracted many charlatans, but then so has every other form of art, including what has been called "classical". One must avoid the tendency to attribute malicious intentions to these people (e.g. that they "have used certain kernels of truth as the means of escaping the extremely difficult problems of our times, restoring to pathology in art and barbarism in politics, solutions of violence whether with a brush or guns").

What I think is closer to the truth is that these movements focus on isolated facets of reality, mistaking them for a whole or a universal essence of the whole. In this connection, I think that from the times of the Impressionists and perhaps before, there has existed a dissatisfaction with the older representational forms of art. It has been felt that they somehow give too superficial a view of reality. There have been many movements, whose main content has been based on the feeling that all the apparently solid aspects of life, as it is seen in common experience, are really not very substantial, and that something else very different lies beneath this. Naturally, in doing this, they have tended to stress what is chaotic, meaningless, deceptive, etc. There is even a grain of truth in such a vision. Consider, for example, the Surrealists, who tend to suggest that behind all the common and reassuring features of everyday life, is something absurd. They do this either by an abnormal increase of irrelevant detail in too uniform and clear a light, or else by taking a picture which is very realistic and everyday in each of its parts, and yet which adds up to something frightening or absurd. In doing this, they suggest something that is true about the world that we live in (e.g.

something like the Nazi concentration camps, some of which were covered with slogans like "work enriches life").

Where all these movements go wrong is that their proponents get so attached to the particular facet of truth that they have discovered that they cannot go on to something new, something more positive and creative. But I doubt that there is any plan in their minds to be violent or to do any dirty work. In this respect, they are no different from some of their opposing counterparts, such as the school of Socialist Realism in the Soviet Union. The latter are reflecting an idea common in that country that what they are building has eternal value. As a result, they are disturbed when any work of art suggests the impermanence and doubtful significance of a great deal of what they are doing. They too cannot look at the whole truth, and engage in violence to suppress the other side, and to force a certain style on artists (also with, in all probability, very good intentions).

I should perhaps add here that my first reactions to modern art were almost entirely negative. However, in some respects, I have changed my mind. For example, with regard to Rouault, I first felt that his pictures were very discouraging and depressing. Gradually, I began to see them in a new light. In particular, last year in London, I saw a picture of his, *The Old Clown*, part of the E. G. Robinson collection, I believe. At first, it seemed to be rather a mixed up set of patches of colour. But gradually, it began to take shape. In particular two patches struck my eye, one in the face of the clown and another outside him, which seemed to complement the first. My eye began to move back and forth from one patch to the other, a pulsation was established, and suddenly it ceased, to give way to a remarkable new steady vision which I can best describe as seen in a new dimension. It was not so much that the clown became visible in three dimensions, this was true but only a minor point. The major point is that there seemed to be a flow or a current in which the whole being of the clown poured outward to reveal itself, all his feelings, thoughts and emotions etc., and a counter-flow in which the outside (including the viewer) was drawn into him, to emerge again in the outward flow. It was a very striking experience for me, one that I shall always remember. Whether the artist intended the picture to be seen in this way, I don't know of course, I would be interested in knowing whether it struck anyone else in this way.

The main point that I want to make in the above discussion is that it illustrates how wrong it is to make snap judgements on something new. No doubt there is a great deal of rubbish in any new movement, but there also may be a great deal of value, which has to be understood in a new light. It is also rather dangerous to make very detailed blueprints as to what the new should be. If we do this, how can what is really new come into being? Of course, we do need theoretical analysis of all existing trends, but we also need a continual alertness, a readiness to recognize the new and the unexpected when it occurs, even if it means altering or abandoning our most cherished views at times. This is perhaps a concomitant of not being "violent" in the sense that you describe in your letter. In other words, it is essential that there be real freedom and no effort to impose preconceived ideas on others, or on oneself.

This letter has grown much longer than I had intended and also due to pressure of work, taken much longer in the writing than I had hoped. Perhaps it could be the basis of an article to be published eventually (in a year or so). I should be grateful for your comments.

Awaiting your reply with great interest,
My kindest regards,
David Bohm

Red Wing
Route 2
Minnesota, USA
May 22, 1960

Dear Mr Bohm:

Thank you very much for your good letter of April 24. Be assured that it was appreciated and reread with considerable interest. I wish I could give you a reply that you would think merits receiving yours.

Once more I have the experience, in your letter as in your book, and which I have not mentioned before, how similar many things you say are to what Alfred Korzybski puts down in *Science and Sanity*, which first appeared in 1933. Do you know of him? Have you read him?

Throughout your letter you are expressing your notions of nature as process, but not until page 3 do you directly use the term "process." You show how everything is related to everything else, including the mechanical instrument, with all of which I agree.

But about those instruments which serve as the perceiving intermediary between the man and nature, about that I should like to know more. Perhaps someone has dealt with this, but not in anything I have come across. What are these instruments doing to the men who use them? Constantly there is the effort to sharpen the perception of the instrument for man, but where is the man in all this, what is happening to him? Too often I feel, while reading science, that the author forgets that he is the final super-instrument, that he forgets mankind altogether along with himself, so submerged is he in the instrumental world. Nature is not a machine. Does man rely too much on the instrument, so much that today he permits it to deceive him because he assumes it no longer supplies a chain of causality? I mean, could it be that the instrument, even though it probes more deeply into nature than ever before, can no longer function so thoroughly for man as it did in the older mechanics? In the sense, that what the instrument now reveals demands more of him, of himself. Is this why men are now erecting absolutes from a state of desperation; a dualism in nature – complementaries – a dualism between man and nature – indeterminism?

When I read accounts of the behavior of the single atoms in relations to all the others, I have the impression of something akin to the creative human being. This is an impression I do not get from the kind of nature picture given by the older mechanics, from the little I know about it. It seems as though the very creative depths of nature are now being approached in physics. This could be the sort of event that mistakenly leads to the view of indeterminism by those too much submerged in over-dependence upon instrumental finality. Now that the instrument finds more, could it be that it "tells" less?

Your use of the term "ambiguity" to express your views on what are all aspects of the totality of nature, is completely new to me. I would have to think more on the use of that term. To me everything in nature is one vast process; "Everything is everything else," as I believe Leonardo put it. Thinking and experiencing in that way, it becomes untenable to accept dualisms of any sort such as complementarianism and indeterminism, and as your use of "ambiguity" seems to suggest too.

I must admit ignorance about Existentialism. I once read a small book by Sartre, but it did not interest me to pursue the subject further, even though there certainly were those "kernels" of truth in it. That there exists a "gap" between our existence and what we think about it is certainly not something that we are the first humans to think about. Perhaps we are so taken with this disparity in our century, because we have lost the means by which we could look with some hope of doing what other times did, namely, to narrow the "gap." Perhaps because, in our times, both the artist and the scientist have literally been ejected from nature's womb, so to speak. This has brought on the crisis in both art and science. That is to say, nature no longer responds neatly to the instruments as it did in the old days of localization, and Heisenberg thinks to take his revenge, while Bohr would tell the disparities of the lasting, the absolute brotherhood of complementaries. As for the artists, they have lost the convenience of nature offering ready-made forms for art, forms which always remained there in nature for the artist to return to when things went wrong. It is not by chance that science and art speak of the "laws of chance," that artists and composers make what they call indeterministic and molecular art. Not that many of these ever bother to read science, but it does appear to offer a solution for artists wandering in a land of drought. It frees them from all restraint from everything, even from themselves, becoming a mere vehicle for the arbitrary.

In your "vision of totality," you just about include all other fields. I think we have to make more modest demands. Why not work to bring into a totality view what is already reaching for it? I mean certain arts and sciences. What a force if the truly creative individuals of science and art were to stop inhabiting different planets and join forces, as they sometimes did in the past, first through a comprehension of each other. These two fields seem to contain the solution for a future sanity, because the solutions needed for each are the ones that mankind seems to be searching for. Alone they continue helpless and largely destructive. We have arrived at the end of "living like maggots in a cheese," as Cassius Keyser warned so long ago (p. 655 my *Evolution* book). Man is now on his own; he is in charge of his own evolution, it will no longer come about in the old "natural" way, he must consciously direct it. Survival now depends, it seems to me, upon being creatively oriented whether one is a scientist, artist or whatever.

Then the relation of our existence to what we think about it, and our relation to nature, will cease to be but a mere problem, becoming more centered in the effort of creation. It is in this respect that I believe art holds the deepest significance to the future (see my "Art and Science as Creation," in *Structure* [1958]). Art has had a very special role in man's relation to nature, which is hardly suspected. It is difficult to imagine how various times could have realized their various views of totality, without art. I also have the impression that what the sensitive individual experiences before art, but largely unconsciously, is a totality relation with the rest of nature. Art makes him feel adequate at being one with the universe of nature.

The artist is without anything but himself before nature, without instruments, without mathematics to order, to suggest, to predict, and without a philosophy. He must forever hurl himself anew and whole into the totality of nature. He cannot dare to think beyond where he is in this totality, except in the most extremely tentative way, otherwise he would frustrate the effort to create art.

What you say about the finite and infinite, about that I have to think more. Even though the finite infiniteness of the infinite throws me, I seem to agree with the conclusions you draw from it. Your notion that "there is really no-'thing' in the world," is a wonderful expression of process. But, in agreeing with you, I have to take what seems like the opposite position. You speak of chairs, tables, etc., how heat applied dissolves all into gas, that even then the atoms can be dissolved into other structures, etc. You thus drive the object down the road that science has traveled where indeed objects vanish into everything. You draw only a picture of nature that science sees, a nature put in reverse. This reminds me of Eddington's famous two tables, the one we are all familiar with, and the one of science. As you will recall, Eddington was perturbed by this situation. What happened, he asked, to art, poetry, to that world of perception and experience which vanished at the behest of the scientific table? Dualisms again, the dualism of art and science.

It is not alone the world of the elementary particle that is admirable in nature, but also the fabulous richness of that world which those particles manifest to natural, human-scale perception. This is to speak in terms of your "totality," is it not? It is not that my kind of artist sees nature only as the limitation of the perceived objects, for us too limited things are only "relatively limited," as you put it. It is rather, that we extensionalize this level of nature into a deeper perception of its use to art.

Let me try again. You write about the problem of the historical past and the future. It seems to me that the physicist is one who, in respect to nature, the further he gets into the "future" of nature, the less he cares about the "past" experience of nature. It seems that what he once saw as the reality of nature, which continues to remain a reality of nature in spite of his denial, should be retained in any totality picture of nature which the physicist makes. I have the impression that the physicist is perpetually seeking an *a* or *the* reality, when all we experience on all levels of nature comprises the totality of nature's reality. On such grounds art and science could meet in a mutual concern with nature, and acting as correctives to the picture each makes of nature.

Accordingly, I emphasize the reverse direction of your illustration, in which we regard the atoms and their energies transformed into other relations until we have chairs, tables and human beings. The direction of nature. I am suggesting that this direction of the totality process of nature can continue further, through the agency of man as artist. While the physicist pursues the creative process of nature, even to how nature began, the artist is working at the other end of this process, extending this creative process. The physicist wants to get at the beginning of nature, to the very birth of the atom. While the artist wants to continue with what that atomic world already manifests in its evolution. Think of the evolution of nature evolving new forms, that is, the new art is continuing the evolution of *nature as* art.

About your discussion of "necessity," I am not sure I understand you. I cannot see how we manage to "remove ourselves from the domain of necessity." If we recognize ourselves as potentially possessed of that quality you speak of as infinite, is not that going from one notion of necessity to that of another, that is, we go to the necessity of the infinite attitude? Aside from that question, I agree completely with your notion that human life is a mode of the infinite, a "reflection" of it, as you say, a wholeness in ourselves that is at one with the wholeness of nature. It is not a question of being that, the human being is such, it is only that we frustrate its free operation.

You write that we do not ignore causality, that comprehending it deeply enough "we find its limitation, which leaves room for us to be free." I wonder, however, if the situation of causality is more similar to what happened to the old mechanics of time and space localization. I mean, is it so certain that the attitude of causality has reached a limit any more than considerations of time and space? The latter were not eliminated simply because the simultaneous thing no longer worked. This is to ask, is it possible that the notion of causality could be extensionalized? That what is necessary is a less mechanistic attitude toward determinism, the inheritance from the old mechanics, that it can be replaced by a creative notion of determinism?

I am very, very far from recognizing any process in an artist which is "free, not determined by anything other than itself." I can only imagine it as a delusion induced by opium, the artist managing to forget the determining role of the opium. Artists of our century, by the car load, presume to possess such a freedom but, as you will note in my writing, I find it to be a fictional freedom at the sacrifice of the unlimited freedom which nature opens to the artist, as it always has.

I believe I have shown, at least no one has disputed it as yet, that even an artist like Piet Mondrian, who claimed independence from the sensory world of nature, remained as bound to it as any artist in the past. His suppression had only the effect of making him unconscious of the role of nature in his art. No artist has ever escaped a deterministic relation with nature, whatever the course his art takes, whatever he thinks he is doing.

The paragraph in which you consider a good work of art as an integral whole, which results from expressing the structure of the universe, and say that its significance lies in its relationship to the totality, with that I am completely one with you. You have put in different words, and very good ones, what I have written many times as being the future goal of art. It is possible, however, that when you have read my writings, you will find that we have arrived at similar conclusions about art from different experiences with nature, since you are a physicist.

In the second section of your letter you give a résumé of what you have been saying. You again speak of the problem of past and future. I think I understand you, and agree. But there is one point I would suggest for your consideration, I will use your excellent illustration of the red and blue lights, the red past leaving the implied blue future. It seems to me there is is one other subtle event that takes place. Looking back at the past we see another set of red and blue lights. Only now the blue light represents any section of the past which was once a future. It leaves an area implied for the red which covers the past of this once future area. We now notice that the implied red area has been altered by the future which followed it. I am suggesting, that if the past is always implying, in some way and degree, the newness of the future, the future is always in the process of implying, in some way and degree, a newness to the past. This is to say, that the past is in a constant state of change which increasingly offers the means whereby we can most successfully discern the potentialities for freedom in the future. This seems to fit in with your statement, that "there is no way to say unambiguously what the past really *was* until we know its future," and "that the whole shall form a unity, as in a picture."

The two paragraphs in which you take up determinism and indeterminism, order and disorder, etc. I can follow you verbally, everything you say is reasonable, but I cannot reach the non-verbal level from which I must suppose your notions to be derived. Even what we usually call disorder, is really a connotation of negation concerning some kind of order. How in the world does one ascertain with any semblance of confidence, that a "particular form of determinism becomes irrelevant"? How does one decide whether it's the particular form of the notion of determinism, or whether it is the general notion of determinism, that is irrelevant? I have asked, "How does one distinguish order from disorder?" You repeat the question, but I could not see that you answered it. Why is there always so much obscurity on this issue? I have the impression that faith is posing as fact.

Here is a sentence that I quote from your book, a kind of sentence that I frequently encounter in science literature: "Because of the *random* distribution of the particles an *almost continuous* pressure is produced on the walls" (italics mine; p. 48). The atoms seems to be doing two paradoxical acts. That is, they are acting in a haphazard and also in an "almost" uninterrupted time sequence. This is a kind of plain disorder and, at the same time, an "almost" order. Isn't there something more here than the argument of statistics by which you resolve such a discrepancy?

You state that in art there has "existed a dissatisfaction with the older representational forms of art," "that they give a superficial view of reality." Here the physicist in you has taken art in the nature direction taken by your field. But art is not science–nature. Those who are the foremost of the post-mimetic innovators, for me Monet and still more Cézanne, remained all their lives as the most fervent admirers of the great ones in their past. They did not need to denounce their predecessors in order to feel status, unlike the Dadaists who had to paint a mustache on Mona Lisa. Monet and Cézanne continued from where their predecessors left off, as did Einstein from those before him, thanking them for what they left behind, which enabled them to go further. Dissatisfaction with the whole past is an art disease that made its appearance precisely with the breakdown of mimetics. Artists, that is most of them, having lost the thread of continuity from the past, blamed the past. Then, not knowing where their art was taking them, they supplanted a definitive effort with a new but strange virtue, not-knowing. It is these not-knowers who shed crocodile tears over art's nature reality which science presumably dismembered beyond further use, but who are in fact pleased that this has happened, pleased that science denounces causality and upholds chance, disorder, indeterminism, etc. When the physicist looks back at Newton, and I am reminded of Einstein, or back at any of the great innovators, does he speak of "their superficial reality"? Far from it, as you know. There is an altogether different attitude. Well, there are some artists too who look at their predecessors with affection, not dissatisfaction.

What you express as the relevance of Surrealism as a reflection of the truth of our times, such as the concentration camps with their cruel, absurd slogans of "work enriches life," is, to be sure, one way to look at it, if somewhat strained. I say this as not being a Jew. It seems to me the perversity of Surrealism to say so. Are not such artists perpetrating the very condition of human life that nurtures such bestiality as the barbarism of the Nazi? Is not Surrealism but art's participation in the insanity of our times? Anyway, with equal reason, I do believe, one could say that the Nazis "suggest something that is true about this world," by pointing to the cruel absurdities that Surrealists conjure in art. This reminds me of the Nazi officers, not small fry, who paid that clown Picasso visits of admiration. Why did they leave this communist stooge alone and murder other artists? Surrealism, yes indeed!

Do you think it impossible for artists and art to sink as low as human behavior sinks in other fields? Are artists some rarefied breed of life, that no matter how absurd, their behavior can never be compared to that among the more ordinary mortals? I am not surprised that such as the Surrealists "cannot go on to something more positive and creative," to use your words. They only know destruction; if the Nazis are the truth, indeed the Surrealists are the truth. You must forgive me, I do not feel these things with the sterility of objectivity.

It seems to me that your reaction to the Rouault picture was the result of your experience as a physicist. As I read this paragraph I almost expected you to say something about a break-through about the quantum barricade. Am I being absurd?

You write that you do not feel qualified to deal with the relations between physicist and art. Do you know of a single soul who is? I do not, whether physicist or artist. Until some individuals in both fields do their best to deal with this problem, even if they must stumble and appear ridiculous, there will not appear a qualified person in either field.

In late summer the editor of *The Structurist*, Eli Bornstein, who lives in Canada, will visit me. With your permission I should like to show him your letter on the basis, as you suggest, that you might make this into an article for a future issue.

I should appreciate your comments, particularly where I try to relate our two fields.

With my best wishes,
Charles Biederman

3, Berkshire Road
Bristol 7
Glos., England
June 6, 1960

Dear Mr Biederman:

Thank you very much for your letter, and for your two books plus the three articles in *Structure* which I received. I have read your book on Cézanne, and the three articles in *Structure*, and I have begun to read *Art as the Evolution of Visual Knowledge*. I should prefer to delay a detailed comment on your ideas until I have finished the latter book. For the moment, I shall only say that I find your ideas very interesting and stimulating. In particular, Cézanne's ideas on space as a unity of interpenetrating planes come close to ideas that I am trying to develop on geometry. I am sympathetic to your efforts to develop an art that is not a mimesis of nature, but rather, something that exists and is beautiful in its own right, having the same general laws of process and relationship as are found in nature. However, I wonder if you may not be in danger of dogmatism, when you assert that this is *surely* the main line in the evolution of art, at least for the present. How can you be so sure that the older forms have been completely exhausted, that art has no more role to play, either as a mimesis, or as the expression of new truths concerning nature? After all, you yourself have stressed that the future throws new light on the past, reveals new meanings in it, etc. In other words, I would be ready to consider your proposals as a *possible* line of development that may well turn out to be fruitful. But how can you be so sure that it must be the *only* or the *main* line of development?

Now to get back to your recent letter (of May 22). I am afraid that I did not express my point of view very clearly in my previous letter. I did not intend to dissolve the familiar large-scale world into atoms, electrons, etc. The opposite process is, of course, equally valid, i.e., of going on from atoms to large-scale realities. Indeed, I would stress that there is no way to *reduce* the large scale to nothing but atoms. From the point of view of infinity, it would make no sense to say that the large-scale world is constituted of atoms, the atoms of electrons, protons, etc., the electrons and protons of something still smaller, and so on ad infinitum. In this way, everything would disappear in an infinite regression. Rather, I would say that each thing, each level, makes its own unique contribution to the totality, a contribution that is not *just* a reflection of something else (even though it may in part be such a reflection). Therefore, the large-scale reality with which the artist works is a part of the whole truth, a part having certain qualities, meanings, etc., that cannot be found *in their entirety or even in their essential features*, by going down to atoms, electrons, etc. Atoms, electrons, etc., are neither more nor less real than the large-scale level of our immediate experience. It is only all these levels working together, each making its own irreplaceable contribution, that constitute the totality.

In connection with this problem, I would argue, for example, that the laws of living, thinking matter (e.g. human beings) contain certain features which are very probably not fully deducible from the laws of the constituent atoms. For example, we have studied atoms, electrons, etc., largely in isolation, or in interaction, only a few at a time. Using laws suggested by these studies, we have then made the gratuitous and unproved assumption that if we could only solve the equations for the myriads of atoms constituting the living animal or the human body, we could, in principle, deduce every property, quality, etc., of the latter. But it is much more likely that something is left out of our laws of individual atoms, something that is of not appreciable importance when only a few atoms at a time are involved, but that may be of crucial importance in large masses of highly organized matter, such as is found in living beings. In other words, living things may have new qualities not found in isolated atoms or in small groups of atoms, such that there are new laws which can be expressed only in relation to these new qualities. Similarly, for large-scale nature in general.

It is here that the idea comes in of the "finite in the infinite". Every thing that is in any sense determinate in its existence at all, must be *limited*. For determination (which has the same root as "termination") means limitation, and therefore, finiteness. On the other hand, we have seen that because the full being of each thing is only in its relationship to the totality, every notion of something as finite is in some way, erroneous. Not only does this follow when we dissolve chairs into atoms, electrons, etc., (i.e., chairs are not just simply chairs, but are potentially and actually infinitely more). It also follows when we build atoms, etc., into large-scale objects (i.e., atoms are not only atoms, they are also potentially and actually constituents of large-scale objects, having further qualities and laws that they could not possess as simple atoms). We see then that determination (or equally well, the termination) of each thing within its pre-assigned limits holds only as simplification valid for some purposes, but not for others. It is in this sense that I say there is no "thing" in the world. Nevertheless, if there were no relative determination, there would only be a characterless chaos. Thus, there is no pure infinite either. There is only the finite in the infinite and the infinite in each finite. These are two opposing categories which complement each other, in the sense that either one immediately demands the other. For example, to say that something is determined or limited is to imply immediately that there exists (at least potentially) something beyond these limits. Thus the idea of limitation immediately implies the unlimited, while the idea of unlimited cannot even be expressed, except in relation to its opposite; viz., that which is limited. And when we have two such opposing universal categories, the truth is that both must be asserted of everything, i.e., everything is both finite and infinite, both limited and unlimited.

I found your comments on the image of past and future in terms of red and blue lights very interesting. It is certainly true that the future is always implying some newness about the past, which "increasingly offers the means whereby we can most successfully discern the potentialities for freedom in the future". However, it is here that the notion of ambiguity shows its importance and relevance. In other words, if the past were already something absolutely definite and determined, how could it in any way be changed at all by the future? If we assert that the past is inherently ambiguous as to what it *is*, at least to begin with, then we leave room, logically speaking, to assert further that the future changes the past, in the sense that *future developments remove some of the ambiguity in what the past was.* This means that on the basis of what may have revealed itself up to a certain point in time, the precise and detailed character of any given thing or event is ambiguous. It is this ambiguity which leaves room for some freedom. For only future developments will show fully what this past really was. On the other hand, guided by the erroneous idea that the past was in itself unambiguous in its nature, we may come to the wrong conclusion that there is an iron-bound determinism, leaving no room at all for freedom in the future.

It is this idea of ambiguity that I wish to apply to measurements in the quantum mechanical domain. When one tries to find the exact location of an electron by means of an instrument, one discovers that the instrument participates so intimately in the mode of existence of the electron that the velocity of the latter must be ambiguous. Of course, in the future, this ambiguity may be reduced, when we see what the electron has actually done, but new ambiguities appear, with regard to the still later behaviour of the electron. *We therefore never get rid of all ambiguities.* As we get rid of some, new ones come into their place. In other words, every description of the electron (or of anything else) must contain some ambiguity with regard to what it is, an ambiguity which means that its future is not *fully* determined by its past *alone*.

It is here that we come to the problem of freedom again. I never meant to say that there are creative acts which are not determined by anything else at all. I only meant to say that they are not determined *fully* by anything else. To explain this, recall that the large scale is not determined *fully* by the atoms, but makes some irreplaceable contribution to the totality of law. Similarly a future event is not *fully* determined by its past, and also makes some irreplaceable contribution to the totality of law. Of course, this does not mean that the large scale is *completely independent* of the small scale, or that the future is *completely independent* of its past. Rather, there is mutual interdependence and inter-relationship of everything. But the law of the interdependence is such that any given thing is only partially determined (limited) by what it is not. *The full law of its determination can be expressed only when the thing in question is included.* Thus, it is more like the law of a picture than a mechanical determination *purely* by externals. In other words, it is only in the *whole* that the *full* reasons for each part can be found; and the whole must include all the parts, including even those whose determination is being discussed. It is in this sense that there is freedom, especially with regard to creative acts. No doubt these acts are conditioned, limited and shaped by society, history, the natural environment and everything else. But these external features *by themselves* do not *fully* determine just what will emerge in this act. It is only when we have the whole act completed before us that we can trace all the reasons why it is just as it is, and in doing this, we shall see that it is only in the context of the whole, including the act itself, that reasons can be intelligently ascribed. Thus, to a certain extent, a creative act contributes to its own reasons for being just what it is. It cannot be reduced fully to a mere consequence of external factors. Indeed, external factors frequently interfere with the possibility of a creative act. And to the extent that this is true, freedom is possible only when we have understood these limitations and removed ourselves from their domain.

I think that in your struggle against modern painters who falsely deny that their painting has any degree of external determinism whatsoever, you must be careful not to go to the opposite extreme, and to deny any degree of autonomy whatsoever to the act of creation.

This brings us to the problem of freedom and necessity. First of all let us ask ourselves "What is necessity?" What is necessary is that which cannot be otherwise. But whenever we consider some part or aspect of the universe, we discover that, *considered by itself alone* it could be otherwise. In other words, we cannot see in a chair the full reasons why it is just as it is. In the materials of which the chair was made there were many *possibilities*. That possibility which is now actual was determined by something external to the chair (in this case, the man who made it). From the point of view of the chair alone, it is only a *contingency* that it is just as it is. (Contingency is the opposite of necessity, viz., that which could be otherwise.) We discover in each part of the universe certain relations that are necessary (laws) and certain properties that are contingent. By broadening the part of the universe under consideration, we may show that some things that were contingent in the narrower context are necessary in the broader context. But then we will have introduced new contingent properties in the broader context, which will be necessary in a still broader context and so on without limit.

This above means that in no *finite* context can we ever get rid of contingencies. In other words, necessity and contingency are two opposite categories, which apply to everything (every finite thing). No matter what we consider, some aspects will be necessary and some contingent.

Of course, we could consider the infinite totality of all that there is, was and will be (the cosmos). If we consider this total cosmic process, then there is nothing outside on which it could depend. Its necessity therefore, is obvious. For whatever the cosmos is, it *by definition* cannot be otherwise. Thus, its necessity and being are identical. The category of contingency has here disappeared. Moreover, necessity and freedom have also merged. For nothing outside can exist which could constrain the cosmos to be as it is. We therefore cannot first imagine a cosmos that could be otherwise and then lay down a law which requires it to be what it is (as we could first imagine a man who could drive on either side of the road and then lay down a law from the outside, such as "keep to the right"). When we have stated what the cosmos is, we will have given everything that exists, every relationship, every law, every limit to law, etc.

When we come to any part of the cosmos, we of course know by experience that each thing has, in itself, many possibilities, and that further laws may be needed to explain why it has to be just what it is. Here, we have an interesting application of the concept of ambiguity. Remember that each part of the cosmos has inherent ambiguities in its mode of being, because what it *is* is *fully* determined only in its relation to the whole (which includes its infinite past and future, as well as itself).

But then we come to the question as to whether that part is necessary. This too will be in some degree ambiguous, because we will actually always try to express this necessity in terms of laws relating to something less than the whole cosmos. To the extent that it is ambiguous as to whether a given thing is necessary, it can be regarded as contingent (and of course, vice versa). We therefore come to a deeper point of view than before on this question. We no longer regard something as contingent in one context and necessary in another. Rather, we say that it is both necessary and contingent *in the same context.* We avoid an outright contradiction here by noting that to the extent that its necessity is ambiguous, there is room for its contingency, and vice versa. This is similar to what we do with regard to the wave–particle nature of the electron. To the extent that its particle nature is ambiguous (so that its precise location is not defined) there is room for its wave-like character to exist (in the domain of ambiguity of position).

When you mention in your article in *Structure* that Cézanne saw the universe as a pulsation of colour, you remind me of some of my ideas on the wave–particle character of matter. We must explain why the same thing, e.g. an electron, can in some conditions act like a wave and in others like a particle. Now, I propose to do it by giving up the notion that the electron is a permanent existing particle (like a microscopic copy of a billiard ball). Instead, let us consider a pulsating model of the electron. We shall begin with a continuous background field extending over the whole cosmos. Imagine a wave coming inward in this field towards a certain point in space. At a certain time, it builds up to an intense pulsation, and then begins to disperse. Think now of a series of such pulsations, very close together, arranged to succeed each other in time along a track or an orbit. If the pulses are very close together, the whole series will reproduce all the properties of a particle on the large-scale level (where the spacing between pulses is too small to show up). Yet on the small scale, the electron is very different. The ingoing and outgoing pulses which make up the electron will explain the simultaneous wave–particle properties of the electron. We do this by giving up the notion that the electron is a permanently existing particle. *Instead we suppose that it dissolves away and re-forms out of the background continually.*

In this model, we see the indivisible unity of each thing with the cosmos, since its pulses come in from the whole, and fall back into the whole. The pulses of each particle interpenetrate those of every other. Hence no particles are really separated in their existence. Even the observing apparatus is made of particles with this pulsating structure.

Thus, every measurement contributes to the structure of the observed system. The old idea of an observation which just gives us information about something without participating in the existence of that thing (and without that thing's participating in the existence of the apparatus) has broken down.

Different kinds of particles (electrons, protons, etc.) correspond to different spacing in time between the pulses, and to different forms of the track (various kinds of spirals, etc.). The "creation" of a particle corresponds to starting a new series of pulses, its "destruction" to ending the series. Transformations from one kind of particle to another correspond to a change in the pattern of pulsations.

Here we have an important idea. If we just have one pulse, we cannot tell whether it is an electron, proton or any other kind of particle. It is only the pattern of a long series of pulses that tells us what kind of particle we have. It therefore has no meaning to talk of an electron existing at a given time. An electron is not the kind of thing that can be what it is at a fixed moment in time (nor is anything else for that matter).

Now comes the question of regularity and irregularity, law and lawlessness, that has been causing so much trouble. From the cosmological point of view, every law is merely a contribution to describing what the cosmos is. For example, we may have the law that the electron follows a certain orbit. In term of pulses, this is translated as the statement that when there is an electron, there is a certain kind of pattern of pulses. This pattern may be more or less regular, in the sense that certain features are repeated, with the result that if we know the pulse pattern up to a certain moment of time, we can predict the further pattern. To the extent that the pattern is this way, we have determinism in time, i.e., the conclusion that the future is *only* a reflection of the past, and has nothing basically new in it. On the other hand, we can assume a more general kind of law; viz., that while there is some tendency for the future to repeat the past (with modifications) there is room for something new to come in, something that is not reflected in the past. We can find the reason for this new thing only when we consider the whole pattern in space and time, including the new thing itself. If we consider less than the whole pattern, we may not find the reason. Thus, *in a limited context* the appearance of such a new thing will have no law. For example, every regular pattern of pulses may have in it sudden breaks, alternations, appearances of new elements, etc. In a context less than that of the whole, we can only say of them that they exist in the cosmos, and must be taken into account. But every context of law is actually finite. Therefore in every theory and in every description, there must appear elements which do not fit into any regular pattern of law, but which are just there. Within the context of the laws in question, these elements are contingencies, breaks in a regular pattern, disruptions of the pattern, etc., or just simply irregularities, having no reason that can be found within the framework of law in question. Of course, they may find their reason in a broader framework, but then further new unexplained elements will appear in the broader framework. We never get rid of such elements altogether. Indeed, from the point of view of the ambiguity in the mode of existence of each thing and in each relationship which it must satisfy, it follows that in no context of law can any regular pattern be shown to be absolutely necessary. Therefore, we will never get to a point where there is no room for something new to come in, something that doesn't fit into any given pattern of law.

Every time we have a regularity, we must have some limit to this regularity. Of course, this may later be described as a new and more complicated regularity, applying in a broader context. But then the broader context will have its own irregularities. Wherever there are regularities, there must be irregularities. Wherever there is law, there must be lawlessness. These are opposite and universal categories, which we assert together about everything (like necessity and contingency).

I hope that this letter answers some of your main questions. In my next letter (which I shall write in a few weeks after I return from Holland) I hope to comment on your very interesting ideas on art and its development. I didn't do it at this time, mainly because I need more time to think about your ideas, and to discuss them with several people who are working in art here (including my wife).

As for showing my letters to Eli Bornstein, I shall be very glad if you do so, and I should welcome his comments. However, I cannot guarantee writing an article. At present, it is only a possibility, indeed a contingency. Meanwhile, I am very eagerly awaiting your reply to this letter.

Very sincerely yours,
David Bohm

P.S. A physicist does not discard an older law when a newer and better one is found. He shows that the older law is contained in the newer one as an approximation or a limiting case (as Newton's laws of motion are contained in Einstein's laws as an approximation holding for velocities small compared with those of light). Therefore the attitude is not so different from that which you advocate with regard to art.

Red Wing
Route 2
Minnesota, USA
June 28, 1960

Dear Mr Bohm:

Thank you for your good letter which I found very stimulating.

I think this last letter has improved my understanding, I hope, of what you have been trying to convey. I find many coincidences with what I believe in.

You wonder if dogmatism is leading me to assert that such an art as Structurism comprises the main line of future development. You ask, has the mimetic direction been exhausted completely? The answer is certainly not. I do not claim that it was, but only that the *structural* capabilities of the two older mediums were exhausted as regards further *useful development* of the past image of art. Hence the reason for the creation of such media as photography, film, television, all of which are structurally more extensive and so able to give the mimetic direction a continued evolution. An evolution, by the way, that appears to have an unlimited future, even beyond art proper.

In regard to Structurism as the future for non-camera art. Such decisions, you are well aware, are based on some kind of assumptions. These make the difference between attitudes that adopt a definitive position about the future course of art as new in a new future, in contrast to those who prefer to delve further on the established theme of what has been done in the past, including heretofore untried variants on that theme. The latter characterizes the conservative effort. But man cannot stand still, or circle a point, in art any more than in science. He must either push forward or he will stagnate into regression. Broadly speaking, the new to the new future is achieved through *constantly* increasing experience and knowledge of the reality of human life within the reality of this universe of nature. This is true for everything, for art as for science. Failing in this the very living quality, the impetus to and the sustenance of a significant art disappears, either as viewed against the background of past accomplishments, or in the view of the needs which the future will require. Everything is then replaced by the placid or violent efforts which have no place to go worth going yesterday, today or tomorrow.

Only in the achievement of this living quality, the quality of adequate existence, does the unique youth of each generation find before them the unique life that is really their own. It is this too which permits an old artist, or scientist, to retain the vigour of youth in his work and life. In our times, however, this pursuit of the inescapable new is confronted with a very drastic situation. It is not a relatively simple change to the new as, for instance, from the romantic realism of Delacroix to the more direct realism of Courbet. Rather, there is a change from one kind of art that all the past has centered on, to one that appears extremely different. This change from mimeticism to an art that is an invention of man, is comparable to and appears to be as drastic a change as that from the old localization of time–space to that of the quantum view of nature, from what little I comprehend of the latter. But it has been my experience, and I would not be surprised if it has been yours in physics, that the more one becomes familiar with the details of the change, the drastic element is replaced by experiencing a logical growth from the past to the present.

Why is it, however, that the quantum outlook upon nature, and for something drastic we have no equal, is so readily accepted while the structural view of nature as a creative process for art, is not readily accepted? The physicist has some marked advantages. These lead everyone to accept his innovations even though few care to comprehend them. Nature herself produces certain manifestations on the physicist's observational instruments that are above dispute, as such, aside from interpretations of them. So that to this day, even though no one has seen an atom, all accept the notion. Then there is the obvious, the fantastic and even fearful "practicality" application of science. All this is connected with what I tried to say in a previous letter about man's naive belief in the machine. To the extent that he over-evaluates the machine he does so at the expense of under-evaluating his super-machine, his own organic perception. In this respect it is of interest that the observing machine has lost its old aura of "objectivity" on the particle level of nature and, like the human being, is found to influence the observed with its "subjective" behavior.

When we come to the creative structural view of nature presented by art, in contrast to the atomic structural view presented by science, it cannot be shown with any instruments whatever, not even as the camera instrument can show the old structural view of nature as objects. The new vision of nature can only take place in the human unaided eye. To make things more difficult, an entirely new mode of visual abstraction is required, and I use abstraction in the sense you use it in your book. When one is bereft of this new vision, or is unwilling to submit his vision to its manifestations in art, how is he going to experience the deeper penetration that has been made into nature's significance to human efforts at art? How can one then comprehend that this vision is a continuance of the past's vision, of both nature and art? How is one to experience that the new art does exactly what all the past has done in the evolution of art? How then is one to see that any "new" variation on the past then pales beside the deeper penetration into nature and art, that is open to our times? It is the latter which the past offers us, not variations on itself. Art lives as does everything in the cosmos, growth or death.

In relation to my views on art, you brought in my emphasis upon the future throwing new light on the past, and which appears to you to contradict my notions of past art. So I must say further just what I meant by that. This "new light" does reveal much that might have been done, and was never done in the past. But it is not our privilege to return to these things, when these things are such as would be comparable to an adult wishing to return to his childhood. What this "light" reveals of the greatest importance is increased knowledge of human behavior in general, enabling us the better to exploit the possibilities of the future. In our times this problem is severely compounded in that we cannot let nature take its course, as the saying goes, but man must now directly make decisions about the future of his evolution. If the future lights up the past, then it does so also in terms of what has never and would never be in the past, but only in the future. Variant possibilities left behind by the past serve our greater understanding of it, but offer no direct course of action for the future. To achieve the last we have to do as the past, create or die.

You correct my mistaken impression that you were reducing everything to atomics and, in so doing, you have also given me a more adequate perspective on the view we both take. We cannot *reduce* everything to atoms, as you say, since that would lead everything vanishing into an "infinite regression." You not only regard all levels of nature as a unity, but also art and science as part of that unity. This is precisely the understanding that I have sought. I am very glad to see you include the artist's level of nature perception as part of the whole truth, especially after what I have read by other scientists who obscure this problem. When you write that atoms, etc., "are neither more nor less real than the large-scale level," this reminds me of a sentence in my "Mondrian and Science" essay, which I mentioned is to be published this year. It reads: "Perhaps the atom is as much a reality, or as much an illusion, as the visible world." This was written in connection with a problem of reality or illusion. It is a great pity I did not have my correspondence with you before I wrote the above mentioned essay. Yet, when you read it, in spite of some of the ignorance I am bound to display about science, I hope you will find something worthwhile in it.

Your next paragraph is wonderful. Where you say that what is revealed to us by the study of relations between a few atoms in isolation, probably leaves out qualities which are present in large-scale nature. This has always seemed incontestable to me, and it is a welcome relief to hear you, a physicist, say it, after all the scientists I have read who speak of the atom as *the* reality. You flatly state that *all* is not atoms. This should ultimately make it possible to coherently distinguish between science and art on their creative levels, and just as important to discern the relations between them. It seems to me that art and science, like all other human activities, form related levels as do the various levels of nature, such as those sought between physics and biology, which latter I do not pretend to comprehend.

I appreciate your effort to further clarify your notion of the finite and infinite, and their mutual qualities. I think you make good sense with what you say there, but I would like to make a comment in reference to it. You write of the terminate as limited, as terminate. In my previous letter I wrote of creative determinism. What I meant by that is that whatever we consider, an atom, particle, tree or human being, everything is potentially of such a character along with its field or environment, that any aspect of its structure has a plurality of possibilities. In this view the determinate, in the sense you speak of it as limited or terminate, occurs only after the event has spilled this way or that, or only after the human has made his creative decision from out of the multiple choices open before him. After that the plurality of choices is manifest again. In this situation what has *become* determinate serves as a guide for future decisions that have to be made before the new set of plurality choices. The finite or determinate, then, becomes a means for demarking, in the form of abstractions, the character of our pursuit of the infinite, unlimited nature of human existence. Accordingly, as you put it, "we have seen that because the full being of each thing is in its relationships to the totality, *every notion of something* [Biederman's italics] as finite is *in some way* [Biederman's italics], erroneous." Are we then faced with "two opposing categories," as you say? On the one hand you show such an acute sensitivity to the mutual relationships in this problem, while on the other hand you feel the need to set up oppositions. I am wondering whether the oppositions are purely verbal or actual.

Could not such a notion as creative determinism supply an explanation, prior to the one of statistics, for the behavior of atoms as explained in the following? "Because we all know atoms to perform all the time a completely disorderly heat motion, which, so to speak, opposes itself to their orderly behavior and does not allow the events that happen between a small number of atoms to enrol themselves according to any recognizable law" (Schrodinger). Otherwise, who is being irrational, atoms or physicists?

There appears to be complete agreement with your further clarifications about the image of past and future. Earlier in this letter I gave some further explanations of my view. Is this in agreement with your position, or do you make a different inference about the future and its effect upon the past?

In your notion of the ever present ambiguity of our experience of nature and, I would add, of ourselves too, as I understand it you are saying, in effect, that we never know everything about anything we experience. This problem not only involves an assumed infinite universe of nature, in which it remains to extensionalize experience and knowledge of it, but is complicated by the fact that the human being fundamentally reacts creatively to each approach to the future. He can react so because each aspect of the future reveals itself as ambiguous, as susceptible to a plurality of possibilities. It seems to me that the ever renewed meeting between the creative being and creative nature is constantly introducing new qualities, to the point that the being himself is constantly becoming new as the universe of nature is constantly appearing new. Once any of the possibilities are acted upon, these possibilities at once impose their unique determinations by virtue of their unique structure. This is not to contend that the creative act is *fully* determined by any one thing or things, but rather, by everything, by what you call the totality. The view of the nature within and without us as a structure of creative determinism, indicates the *kind* of "freedom" open to human life, fundamentally the freedom to be creative. This seems in agreement with your important statement that only when the whole, complete context, which includes the creative act itself, is before us can we ascertain its reasons or, what I would say, *tentatively* ascertain its determinations. And, to the extent that each creative act makes both the human and his vision of the universe constantly new, this relates with your notion that the creative act itself contributes something special to the nature of the whole context being considered. A striking example, I cannot think of a more vivid illustration, is the purely creative work of art, visual or auditory.

The above paragraph is also meant as an answer to your warning that I not go to the opposite extreme of those artists who reject all determination, by denying any degree of autonomy to the act of creation. I am extensionalizing, humanizing mechanistic determinism to a view of determinism as creative, regarding the creative act as the epitome of freedom open to human nature within environmental nature. What is freedom? Is there freedom? Is not this a problem man's stupidity has gotten himself into? Is there not something better – to be creative?

What you write regarding necessity and freedom, that is, that any aspect of the universe considered alone could be otherwise, seems at one with the attitude of creative determinism. But again you bring in dualisms, this time between necessity and contingency as oppositions. Yet you then draw a picture of the universe totality, in which both contingency and freedom merge with necessity, which is agreed to. I do not see, however, the *necessity* for considering them as oppositions. It seems to me you begin with some things you call oppositions, then whittle these characteristics away until they are non-existent, then you merge them in a happy embrace. In every abstraction we make we inevitably leave out a multitude of structural aspects, whether we want to or not, so that all abstractions are accompanied by contingencies. The latter, sooner or later, submit to the form of necessity. We have good reason to believe future contingencies will respond in the same way. Ultimately, it is necessity which appears fundamental to all our experience of structure, not contingency. Contingency exists only in our abstractions, not in what we abstract from. If, as you write, necessity and contingency must occupy every context, then this is because we never know what the totality consists of.

When I first read your notion of a pulsating model of the electron, I thought of Einstein who, on arriving at a solution for a problem, remarked that it was so "beautiful" he hoped it were "true." That was my feeling towards your solution. It is worthy of being true to nature. You seem to have eliminated the stop-gap necessity of complementaries along with dualisms, as concerns the wave–particle problem, supplanting it with what you call an "indivisible unity." I wish I were capable of responding further to your solution, but all I am capable of is to admire it. I wish your model all possible success. Let me know how it develops.

Whenever I read that the observing instrument participates in the structural determination of the observed, as in your letter, I have the feeling that something extremely important is taking place in this situation, as though there is some immense and simple secret contained in it. The striking thing is that man created observing instruments to avoid the subjectivism of his own being, but now finds that his instruments exert a subjective influence on nature too!

You write that the electron, or anything else, is more than what it is at any fixed time. This says something incomparably more meaningful to me than all the talk about indeterminism. In a way, it says what the indeterminists think they are saying, but fail to say. I feel the same way about your method of distinguishing particles according to the kind of rhythm and pattern of the pulse, the significance of transformations in the pulse, etc. All this seems to have the character of our experience of nature – structure as creation.

Your explanation of regularity and irregularity, law and lawlessness, is clear and of a piece with everything else you say. One question to this, the same old one I have been making. Since all contingencies eventually resolve into law, even though this evokes newer contingencies, then why must one ever apply the connotation of lawlessness to any contingency? Why is it necessary to preface this situation with a schizophrenic structure of law and lawlessness? In my next article in *Structure*, "Symmetry in nature and Art," I consider the problem of the kind of regularity that symmetry evokes in nature and art. I hope to get you a copy. There you will see what you call irregularity is considered a characteristic of the regularity manifest by nature on all levels.

When you give me your criticism of my art and theories, and which you say will be the result of consulting others and your wife, will you do this? Let me know what specific artists your group admires, especially after Courbet. With that I could better comprehend whatever criticisms you make, and better answer them. Does your wife have snapshots of her work or that of the other artists, which I could see? In any case, I look forward to what you will have to say, with the greatest interest.

Meanwhile may I thank you for your patience with my efforts to understand your work.

My best wishes,
Charles Biederman

II

experience, so that one removes the "and" and writes contingency-experience for the same reason that space-time was hyphenated. This attitude does not prevent you from "showing in a non-trivial way how the two factors interweave in a pattern of structure to make one.."etc. They not only interweave, they are interwoven to begin with.

Your example of the painting in relation to the above, interests me very much. I believe I understand you, but I'm not sure its your understanding. So, could you illustrate it using geometric forms? For instance, suppose we make a painting with two lines so→ 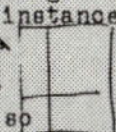 and we assume that all this is properly organized, the parts to the whole. Even this limitation of factors, has a multitude of contingencies and so possibilities. I can change the shape of the canvas quantitatively, the width of the two lines, the right angle relation of the two lines, etc., etc. Now would you call a contingency that has to be removed, in order to preserve the proper structure of the whole, so→ 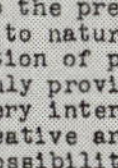 ? It so happens that the most difficult problem in Structurist art, is to discern just those pam particular possibilities that will achieve the development, that will sustain the evolution of an art of pure creation, as against indulgence in arbitary possibilities.

Your discussion of the possible and actual interests me a great deal. Yes, we can agree that "possibility is not just a subjective judgement." But I would do so in a context you would disagree with. That is, possibility is only inherent in the interaction between a human being and nature, and not in extra-human nature as such. I mean this in the same way that the act of abstraction is not inherent in nature, but only due to the presence of humans who are able to relate themselves to nature with abstractions. Hence, the objectivity of the notion of possibility has its base in human action, nature only providing the field for such action. One can make a very interesting distinction between the mimetic and purely creative artists relations to nature, where we find the term "possibilities" has two greatly different contexts, and so meanings. The mimetic artist's possibilities ARE predominately determined by his capacity to discern the already present actualities perceivable in optical nature. Whereas, the purely creative artist's possibilities rest not only on the capacity to discern what the mimetic artist perceives in nature. He takes optical perception further to center his focus on the process character of nature's structure. Thus he is able to discern possibilities that are potential to nature but are not actual in nature, and which would never become actualities without the human artist. The mimetic artist discovers and largely records what nature has already actualized; the non-mimetic artist is a discoverer of hitherto non-actualized possibilities, which only human nature can actualize.

From the above you can see the importance I attach to the

CREATIVE DETERMINATION

3, Berkshire Road
Bristol 7
Glos., England
August 1, 1960

Dear Mr Biederman:

I am sorry that my answer to your letter has been delayed so long. I have been away, and since returning, I have been busy. However, I hope now to make a few comments on your very interesting letter.

First of all, before discussing questions of art, I would like to continue on certain general questions of principle, because I think that these questions are very deeply involved in what I would like to say about your ideas on art.

In your latest letter, you expressed views on creation which come very close to my own. Your idea of creative determinism (which I would call "creative termination", or as a compromise, "creative determination") is more or less the transcription into the artistic sphere of my idea of the cosmic process forming itself and thus creating space, time, matter, movement, etc.

The nature of each theory can only be understood in terms of the law of the pattern or structure, which represents its part in the whole process. Ordinary things represent situations in which the pattern repeats itself with modifications, but the creative aspects of the process are emphasized at certain critical points, where the non-repetitive character is of crucial importance.

I would like to add that the notion of creative determination should apply not only in science and art, but even more, in human relations generally. For above everything else, the human being is so complex and dynamic in his structure that his very character is determined in the interplay of relationships with the whole universe and especially with other human beings. The ideal of human behaviour should be that every moment of relationship be creative. Naturally, this ideal is difficult to achieve, since in a great many spheres (industry, government, politics, education, etc.) the relationships necessarily must have some degree of repetitiveness and fixed character. But at least some moments ought to be creative. Our mode of life today is however such as to make this almost impossible. People are not really related to each other freely enough to allow something new and creative to emerge between them. This lack of freedom is often referred to as experienced as a "difficulty of communication". But the notion of lack of creativeness is, I think, a more accurate way of describing the trouble. The idea of communication presupposes that each man has something in him, an idea, an emotion, etc., which he wishes to communicate (share in common) with other people. If this were all, then we would really have little in common, because the content of each man's mind would be his own only, and there would be no basis for communicating it. But in a genuinely creative relation, the new theory that comes into existence comes about through the relationship, and therefore, it must be something "between", something shared. There is now no mystery in communication, since what is to be common was in the first place produced "between" the two people.

In art or science, man may have a creative relationship to other people by way of his material or ideas. However, in current society, it is hard to have a direct creative relation to other people. The indirect relation by science, art, literature, etc., is the most that can generally be expected for the present period at least. It is here that science and art may play an essential role in the evolution of a new mode of existence for humanity. For it is in these fields that man may first obtain a clear idea of what creativity really is; and from these fields, this idea may spread to human relations in general. When human relations generally tend to be based on creative determination, we may have hope for a happier, more stable, and yet dynamic and ever changing society, in which each individual may find room for his own peculiar capacities in his contribution to a harmonious and many-sided whole.

I would say that the work of art or science is not only a creation of the man who does it, but that its function in the viewer or the student is to get in movement a similar creative process. It is in this way that something is "communicated". But what is communicated is not something belonging only to the man who did the job. Rather, it is also a potentiality of the man who views it, so that a creative person helps to reveal to others their own potential for creativity. (This is a quotation from John Berger, who frequently writes in *The New Statesman.* Have you ever read him?) In the same way a non-creative person (i.e., one who tends to repeat or imitate) reveals to others their potential for non-creativeness, and thus serves to dampen and depress them. Here is a case of how the character of human beings is formed in their relationships. But I believe that in the long run, this kind of formation of character goes much further, because, as I said before, what counts is not only the creativity of a man by the intermediary of inanimate material, but even more, the direct mutual creativity of two or more people who have the right relationships. And here, something new can come into being in the character of each person, something which must originate in such a relation between people, but which can creatively determine new characteristics that never existed before. As you stated in your letter, these new characteristics will then be the basis for still further development.

I believe that the above ideas on creativity are very close to yours. Is this so?

As I see the problem now, our main source of disagreement is on the question of law and lawlessness, regularity and irregularity, necessity and contingency, etc. We agree that in every finite view of the world, which must be partial, incomplete, an abstraction, there will appear limits to law, regularity and necessity. However, it seems to me that you regard these limits as *nothing more than* a result of our necessarily "incomplete knowledge, at any particular stage in our development". In other words, you would maintain that objectively speaking, there cannot exist such characteristics as lawlessness, irregularity and contingency, these being merely subjective characteristics, which we erroneously ascribe to reality as a whole.

The difficulty in the above point of view is that in effect it also denies reality to time, process and creativity itself, while it admits the reality only of the infinite and the eternal. To see why this is so, recall that in the point of view that we both agree with, the past is regarded as being, in a certain sense, incomplete. At any given moment, the past is "what has been". It is terminated, hence determinate. It is also of course the basis of the future, but considered in itself, it defines only possibilities, which are actually determined "creatively" by the future when it comes. After this future becomes the "has been" of a still later future, we can then see the laws which made just this creative choice necessary. At this stage, you would perhaps argue that the necessity was there all the time and that our lack of knowledge was what hid it from us in the beginning. But here it is important to notice that what was originally hidden from us was just the future itself; viz., the creative choice which did not follow from anything that "had been" up to the moment in question. If you assert that this creative choice was actually necessary *from the beginning*, you deny that it really made an essential and irreducible contribution to what the totality *is*; and you fall back into ordinary mechanical determinism, in which the future is *nothing but* a reflection of the past. On the other hand, if you assert that it is necessary from the point of view of eternity and infinity, and if you state that its lack of necessity from the point of view of the development up to a given moment in time is nothing more than a subjective error due to our lack of knowledge, you also deny that the act of creation makes an essential and irreducible contribution to the totality. This time, however, you jump to the opposite extreme of *teleological determinism*, in which the eternal and infinite end is the only reality while the temporal and finite intermediate means has no degree of autonomy in its role whatsoever. This latter position is essentially that which has been held by many mystics, especially in the East, where the reality of time is often denied. The former position (mechanical determinism), which has been more common in the West, also in effect denied the reality of time, because it leaves no room for creative determination in the future,

It seems to me that if you want to say that the creative act is a real determination, and not just a reflection of some past beginning or future end, you have to admit that the incompleteness in what "has been" up to any given moment is just as real and objective as is the completeness in eternity. It is true that at a given moment, we do not know what is to come. What is fundamental, however, is not our lack of knowledge, but rather, that in what "has been" up to this moment, there simply do not exist *all* the factors that determine what is to come *completely*. This fact is objective and not just the result of our lack of complete knowledge. Rather it is the other way round. Our lack of knowledge results *at least in part* from the incompleteness in the being of what has been. Since our knowledge is based on what has been, it too must be incomplete for this reason alone, as well as for other reasons (viz., that we do not even know the whole of the past).

From the point of view described above, the creative process is real. For at each moment, certain aspects of the world are incompletely determined (ambiguous, and having a range of possibilities). It is this indetermination which leaves room for the creative process to make an irreducible contribution; viz., to terminate a range of possibilities and to determine one of them as actually existent. But the creative process has another equally necessary side; *it opens up a new set of possibilities.* Hence, it is not *just* determination. It is determination on its past side only, but it is the opposite of determination on its future side. These possibilities are at the moment of their being opened up, not fully determined by anything that "has been" up to that moment; their full determination requires a later creative act. To say that everything is fully determined in the eternal totality is correct; but it leaves out the essential fact that there always exists a range of possibilities that are not fully determined in the totality of what has been up to a given moment of time.

This brings us to the question of necessity and contingency, which is very intimately related to the question of actuality and possibility. As I said in my previous letter, it is of course true that whatever the cosmos in its aspect of eternal totality is, it cannot be otherwise, for the simple reason that there is and can be only one such eternal totality. Thus, the fundamental starting point is that the cosmos is *necessarily* as it is. But this statement is, *taken by itself alone* quite empty. For in order to go further, and to give some content to our ideas about the cosmos, we must introduce the notion of parts of the cosmic process existing in space and time. In any part, it is *necessary* that there be contingency. For the basic defining quality of a part is that it is *not the totality* and yet inherently related to this totality. As a result, it must be dependent on what it is not. This is just the definition of a contingency; viz., that its necessity does not follow *fully* from what it is. Note that the contingency of such a part is not just a result of our lack of knowledge of the factors that make this part necessarily what it is. For in order to be a part, something must be such that the full reason for it is not itself. As a result, *even if we knew everything, the above quality of contingency of each part would necessarily remain the same.* A correct knowledge of the factors which made this part necessary would show at the same time that this part, considered in itself, is *necessarily contingent.* And we could not eliminate the concept of contingency except by denying the reality of space, time, partial aspects of a process, etc., in favour of that of the infinite and eternal totality alone.

In your previous letter, you objected to my setting up the opposition of two categories, such as necessity and contingency, and then allowing the opposition to vanish in a happy union. But this is just the essential point. The truth about the cosmos is to be asserted by first asserting its unity, by then asserting its duality in terms of opposing categories, and then showing the interwoven unity of two duals. If we did not assert duality, then all that we could say about unity would be trivial and empty, the mere repetition of the words "The infinite and eternal cosmic process is one." By then saying that this process is also dual, we set ourselves the problem of showing in a non-trivial way how the two opposing categories interweave in a pattern of structure to make one, and thus we obtain a description that is potentially capable of being rich and full of content.

With regard to necessity and contingency, we see that to make a non-trivial statement about necessity, we must assert the contingency of partial moments of the cosmic process. Here, I must emphasize again that it is not just our lack of knowledge of the totality that is involved. Rather, what is at stake is the question of the mode of being of a partial moment. In other words, in order that partial moments shall really exist, it is necessary that they shall contain partial chains of necessary relationship, which come to an end. It is in this way, that one can give a definite meaning to the concept of a partial moment. If no such partial chains of necessary relationship existed, there would be no justification at all in our procedure of distinguishing between different moments. Thus, we assert the *essential and necessary contingency* which makes a distinction between different moments meaningful and correct.

But now we must come to the "happy union" of contingency and necessity. For the limits on each particular chain of necessary relationship are seen to follow necessarily from the relation of this partial chain to the totality (including itself). Of course, it may be that the limit on a given chain of necessity will follow from a law. More generally, however, we must take into account the fact that each existent thing and action contributes to the totality of law in its own special way, that is not fully reducible to that of anything else, or of any general category of things. As a result, a given form of necessity can be limited not only by more and more nearly universal forms of necessity *but also by necessity arising from the existence of the particular and the singular*. (It is here that the human creative act can play a key role.)

We see then that necessity and contingency interweave and that each calls for the other. The mere statement of necessity is empty; there must be contingency of partial moments in order that the concept of a necessary relationship can be formulated in a non-trivial way. That is just the way science, art, and every other form of human endeavour go. After we recognize the general principle of necessity, we distinguish the partial elements in our problems by recognizing their contingency and then we obtain meaningful laws by recognizing the necessary relationship that must be expressed in terms of these contingencies. If there were no meaning to contingency, it would be no achievement to find a law showing a necessary relationship in the contingencies. It is rather like the relationship between a picture and parts of it (e.g., spots of paint). Each part in itself is a contingency, yet if it is a good picture, each part is seen to be necessary in view of the whole. If however each part did not in itself have some contingent aspects (i.e., that were not necessary on the basis of anything in itself alone) then there would be no room for the picture to achieve something by removing this contingency and relating each part to the whole. In other words, the necessary condition for an overall necessity and unity is that each part shall have a genuine contingency that can be removed by the picture as a whole. But this creation and removal of contingency is not just a game. Rather, it is the essence of the process by which the picture can really be a picture.

More generally, then, necessity and contingency must be interwoven in the correct description of anything from any standpoint whatsoever. It is true that what is necessary in one context may be contingent in another and vice versa, in the way I discussed in previous letters. But this does not mean that only the context of the infinite and eternal totality, with its absolute necessity of everything, is correct. Rather, it means that the interchange of the roles of necessity and contingency in our thought processes reflects the interchange of roles that takes place in reality. For example, what is a contingency in terms of all that exists up to a given moment of time may be a necessity in terms of the later moment (otherwise, as I have pointed out, we must deny the reality of time). Therefore, when we adopt two contexts (i.e., that of one moment of time and that of a later one) this is not just the result of ignorance or arbitrary choice, but rather, it reflects that each process has *in itself* many equally valid contexts. One of the jobs in correct thinking is to try to reflect this by adopting a set of contexts existing in the process itself. Of course, it may often happen that our choice of contexts is artificial, and not a true reflection of the real contexts in the process. But then this is only a mistake in our thinking. (Just as we may mistake one object for another, or use inadequate categories, etc., etc.)

We come then to the idea that nature has its own contexts, and that it is our job to find what these really are. We can clarify this problem a bit further by returning to the problem of possibility and actuality.

Now from the point of view of the infinite and eternal totality, the category of possibility is meaningless. For whatever this totality actually *is*, it has no possibility for being otherwise. The category of possibility applies only to partial moments. But here, I think we both agree, that possibility is not *just* a subjective judgement on our own part, which we ascribe to things because we do not know their full and concrete actuality. Rather, there is some sense in which possibility really belongs to a partial moment, so that when we think of its possibilities, we are reflecting some aspect of reality. But here, one must stress that *even the possibilities that are not actually realized are in some sense real.* In other words, if we wait until we have some approximation to the eternal totality, we will see that only certain possibilities have actualized themselves. Nevertheless, for a full understanding of the process, we cannot ignore the possibilities that never become actual, and that in general can no longer even be actualized at all.

The above would lead us to consider the notion of real possibilities or *actual possibilities.* Even if none of a thing's possibilities are ever actually realized, we can ask if they are really a possibility (i.e., it is really a possibility for an acorn to become an oak tree, but not to become a lizard. This, we can say with meaning, even if the acorn is never planted).

I shall now try to give a more precise account of the relationship of possibility and actuality from a general point of view. In doing this I shall first stress the notion of *possibility as a relationship*. In other words, one will not talk about pure and abstract possibility, but rather about possibility in view of such and such a fact. Let us consider a given moment in time and all its possibilities. Each relationship to the totality (including itself) may narrow down its possibilities. The totality of all relationships would narrow them down to actuality. Thus, each moment is in effect framed by a series of penumbras (or haloes) of possibilities, one within the other, and gradually converging down to the actual character of the moment in question. Each set of possibilities is a reflection of a certain partial relationship of that moment to the totality (viz., it tells us how the moment is limited by the partial relationship in question). The limitation is not just the result of our lack of complete knowledge concerning actuality. Even if we knew the totality of what is, was, and will be actual, the full statement of what each moment really *is* would lead us through all its relationships, and therefore through an implicit definition of its series of penumbras of possibility. In this way, all the possibilities of each moment, even those never actually realized, are seen to be essential aspects of what that moment actually *is*, because they reflect its relationships to the totality.

The above implies a union of the opposing categories of actuality and possibility. At a naive level, one might be tempted to say that only the actual is real (at least only what is actual in the infinite and eternal totality). But then to give the category of actuality a meaning, we must introduce the category of the possible. Again at the same naive level, one might argue that possibilities are unreal, and are ascribed to something only because we do not know what will emerge actually in the infinite and eternal totality. This is similar to the view that contingency represents our lack of knowledge, and that only necessity in the totality is real. Both views are inadequate, because they do not take into account the reality of the partial moments.

What we have done here is to first introduce the category of the actual. We then oppose this to the category of the possible. Finally we are led to a new description, by uniting the two categories in a concrete way; viz., by coming back to the actual as a limit of the series of penumbras of possibility, and by representing possibilities as relationships in the actual cosmic process (i.e., each range of possibilities is determined by a relationship that actually exists in the process). Once again this procedure of introducing duality and then bringing the duals together in a "happy union" is seen to be necessary to describe the world as it is.

We may usefully compare the above conclusions with those concerning necessity and contingency. After introducing contingency as the opposite of necessity, we showed that just as there are "actual possibilities", there are "necessary contingencies". This is an example of the interweaving that is characteristic of opposing categories.

September 23, 1960

Since starting this letter, I have become very busy and could not finish it. Now I must go away to Israel for about six weeks, but after I return, I expect to continue. Meanwhile I shall be very interested in your comments on what I have sent to you.

We have been having many discussions on your books and articles (which have now been lent to some artists). When I get back I can let you know about this discussion. Also while we were in Holland, my wife and I had an interesting discussion with Carl Visser, whose work was then on exhibition at the Rijksmuseum in Amsterdam. I shall let you know about this later too.

Incidentally, it appears that Pasmore is an exponent of the Structurist school. We have seen some of his work recently. Would you agree with this?

As for my wife, she is still studying painting. Unfortunately we have no photographs of her work. She has done some sculpture which is what I may call an effort to express the meaning of certain mathematical ideas that I explained to her. Right now she is intrigued with what can be done with a pattern of interweaving opposite spirals in painting. She also does what you would call "mimetic painting".

I would appreciate hearing from you soon.

Very sincerely yours,
David Bohm

Red Wing
Route 2
Minnesota, USA
October 3, 1960

Dear Mr Bohm:

I am glad to receive another provocative letter from you, and I appreciate the effort you have put into it. I am also glad that in a great many ways we pursue a mutually related path in respect to understanding this very difficult problem of creation, whether in nature or human nature.

Your revision of my expression "creative determinism" to that of "creative termination" or "creative determination," I find is so excellent that I intend to adopt them. I am also going to insert these changes in an essay I wrote some three years ago called "Art and Freedom," in which I take up the problem of "determinism" in art.

Certainly you are very right indeed to insist that creative determination applies to human relations generally, as in science and art. It was to that thought that I devoted my closing paragraphs in "Art and Science as Creation." I have the horrible premonition that if there is not to be a more general awareness of creative determination, the consequence will be to destroy such a pursuit even in science and art, leaving us with a "Brave New World."

At present, as you point out, there is lacking the freedom so essential for a creative amalgamation to occur between people. I can feel this only too well, since I have lived in a state of complete isolation here for the past ten years. Even the scientific and institutional popularization of the term "communicate," in place of such terms as "converse" that were once used, indicates the current problem. For the former means only to "give," the latter to "do with," all of which coincides exactly with the criticisms you make of current communications, criticisms which I was not aware of as you present them. You say, however, that it is not a lack of freedom, but a lack of creativeness that makes for difficulty of creatively conversing with one another. I would not say it was a lack of creativeness, but a fear of being so especially in our times, because of the very high price one must pay to be creative. Conformity is the great goal imposed on both sides of the Iron Curtain. Hence it follows that the general effort everywhere is one of "communication," of giving directives or inducing people to accept them. To "do with" or to achieve what you call "between," in which both are creative participants, is indeed rare. This cleavage between people is reflected in the cleavage between man and nature. The latter cleavage, I feel, is playing a major role in coloring the kind of nature theories that are dominant today in both art and science. Man and nature disappear, and in their place is something floating called science, art, etc.

Your hope that science and art could serve as the means for achieving the new mode of creatively oriented existence for humankind, is just what I tried to say in a previous letter to you. So we are agreed on a place for hope! But so far the facts are that throughout this century certain kinds of science and scientific *attitudes*, have been determining human relations, wielded by the hands of politicians and other institutional mentalities. The science of psychiatry largely resolves the hot war of the individual into a cold war and solves nothing. It is to what is left of such lives that most art addresses itself, a form of distraction. Until art, as it did in the past, again becomes a sane participant in determining the orientation of human life, we will remain in trouble. I do not think that science alone, as it seems to me the last 50 years have indicated, can solve our difficulties. Nor do I believe that art can do so alone. Art and science have in them qualities that could neutralize the destructive ones inherent in each field, since both have the potentiality for both good and evil.

Yes, I am familiar with John Berger, since I have been reading *The New Statesman* these past two years. Although his views on art are very far from my own, he has occasionally written things I agree with, such as his remarks on that monstrosity Gabo erected in front of some department store in Amsterdam, and which you probably saw on your visit there. The quotation you give from him is to be agreed with; where I would disagree is what he would like to see "communicated" as art. His political views give a nineteenth-century hangover to his notions of art. However, the way in which you go on to discuss the sustenance of creative relations, is more to the point.

Apparently I have not made myself understood as regards creative abstraction before a future as it unrolls into present experience, and then into the past. But I am glad you did misunderstand me, for thus you brought out certain criticisms of the general view I hold with you and, in dispensing with them, brought out some factors of which I was not aware. Certainly, if I thought as you assume, I would indeed be but perpetuating the same old mechanical determinism.

When creative abstractions are made before a future as it comes into experience, the necessity that I take to be involved *is not* the particular abstractions made since, as I tried to make clear, the possibilities are multiple not singular. (Although I think if we could know all the possibilities of abstraction open to us before each piece of future, and the consequences from each, we would find one abstraction that was superior to all other possible abstractions, with which you will not agree.) However, once we make a creative decision among the possible abstractions we discern, necessity lies in the fact that whatever the abstraction made they will be like past ones, in the sense that they are made by the *particular determinant structure* of human life out of the *particular* determinant structure of nature. Therefore, it is not a case of particular determination but of a general process of determinant structure that remains in force whatever the creative abstraction we may make. In this sense abstraction is indeed creative; it does bring in a new contribution even, although we may not always see it, even *of the very notion of determination* as such. The temporal to the contrary, is not denied, for every creative act, every passing moment is potentially open to new creation, creation itself is constantly undergoing change along with everything related with it. This is based on an attitude of non-identity where anything is not the same as it was or will be or will ever be after that. Here there is not the possibility of mere reflection of "some past beginning or future end," to use your excellent expressions. In fact, such events are impossible. When I wrote that the "creative act is fully determined . . . by everything, by . . . the totality," I meant to express the view that all was a determinate process, that the character of the totality was determinate. I was not assuming a "future end" that would impose itself upon each future that we would experience. I also meant that not just any creative act was possible, or that all were of equal value, among the manifest possibilities open to abstraction at any one time.

While I'm on the subject, I'd like to say something about what you have labelled totality, to which I invite your criticism. In your previous letter you said that whatever we deal with, it must be finite. This is invariably the case unless we are inflicted with a case of mystics. You then carry your thought further, stating that it is also necessary to regard the totality itself as "self-limiting" since to regard it as unlimited "would be nothing at all." A little further along in your letter you say that the totality "includes all that there is, was and will be." From this I infer the totality to be some closed system, an absolute system that can be no more nor less than what it is *all the time.* Here it seems to me that the creative quality, and the temporal one upon which it depends for its reality, are both flatly denied. The totality being what it always was, is unaffected by the passage of time, where one cannot distinguish a past from a future and, as you well show in your letter, to ignore these factors is to deny the possibility of creation. The whole becomes the opposite of its parts, a dualism. There is another possibility from what I called, in an earlier letter, the "finite, infiniteness of the infinite" (a contradiction in terms), but which seems to give substance to the notion of the totality as actually unlimited.

If we extrapolate from the views we have been discussing, and to which we mutually hold, it seems another view of nature is possible. One in which the "incompleteness" we experience before the "has been" and the "will be," is also true of the cosmic totality itself. That is, the cosmos itself is undergoing a constant state of growth, and is not any more "self-limited" than any of the finite experiences we have. In this view, if one could at this very moment possess the total experience and knowledge of the totality, one would only experience that moment of it, but not the totality. For, the very next moment the totality would evolve to another totality, just as do our finite experiences. There is a structural correspondence here between our creative abstractions and experiences, and the cosmos as creation. In this view, it is not the totality of the universe at any one time that *is infinite, it is only the creative process which permeates the structure of the cosmos which is infinite.* It is for this reason, as I hope you will soon see in my Mondrian and science essay (the title was changed at the last minute to "The Real and the Mystic in Art and Science," supposed to appear this month), that I wished to make the notion of creation fundamental to further comprehension of nature's structure. This is certainly as true of science and so far you alone, to my knowledge, have confirmed this. Are there others in your field who do so? This reminds me, I would appreciate any suggestions about books, ones that are possible for me to understand. I have no access to a library except the local one in R.W. [Red Wing], which is hopeless for my interests.

Your remarks on necessity and contingency. Here too you begin by positing a totality as closed and absolute. Nevertheless I agree with your further remarks about positing a cosmic process as made up of parts, and which parts are not the totality yet related to it, and that each part "must be dependent on what it is not." You then go on to show that this situation indicates the contingent character of any part, because "the full reason for it is not in itself." I am in full agreement, but there is a difficulty. Is it a semantic one? I do not know for sure. I don't see contingency as an attribute of anything outside ourselves, to me it appears as a state of our consciousness between the "has been" and the "will be." That is, we are conscious that all abstractions about the partial are themselves partial abstractions, and all statements about anything should be followed by an etcetera, to indicate that which has inescapably been left out, and which further experience will make a little more evident to us (we hope). Contingency exists by virtue of the kind of structural abstracting relationship of the human being with nature, except that human life is nature. Contingency consciousness on the part of the human, of *his* experiences *of* nature, not nature as contingent, is of the very essence that makes for a creative consciousness of nature's multi-ordinal facets of determination. Contingency is a fact of experience, not of the actuality experienced. There would not be any contingencies if there were not humans to experience them. My objection is to projecting onto nature, what belongs only to human nature. Otherwise, it seems to me, contingency is applied anthropomorphically.

Duality. It seems to me you assume necessity and contingency as oppositions a priori, you do not assure me of its actuality. To say that one needs dualisms to do this or that with the analysis of nature, is no assurance that there are dualisms in actuality by which one can do this or that with analysis of nature. What does the notion of contingency do for that of necessity? How does one operate with the other? Contingency awareness extensionalizes our awareness of the finite in its multi-relations to every other part of the whole cosmos. There is not necessity to my left opposing contingency to my right, which I then bring into a happy harmony. They are together all the time in experience, so that one removes the "and" and writes contingency–experience for the same reason that space–time was hyphenated. This attitude does not prevent you from "showing in a non-trivial way how the two factors interweave in a pattern of structure to make one," etc. They not only interweave, they are interwoven to begin with.

Your example of the painting in relation to the above interests me very much. I believe I understand you, but I'm not sure it's your understanding. So, could you illustrate it using geometric forms? For instance, suppose we make a painting with two lines so and we assume that all this is properly organized, the parts to the whole. Even this limitation of factors has a multitude of contingencies and so possibilities. I can change the shape of the canvas quantitatively, the width of the two lines, the right angle relation of the two lines, etc., etc. Now would you call a contingency that has to be removed, in order to preserve the proper structure of the whole, so? It so happens that the most difficult problem in Structurist art is to discern just those particular possibilities that will achieve the development, that will *sustain* the evolution of an art of pure creation, as against indulgence in arbitrary possibilities.

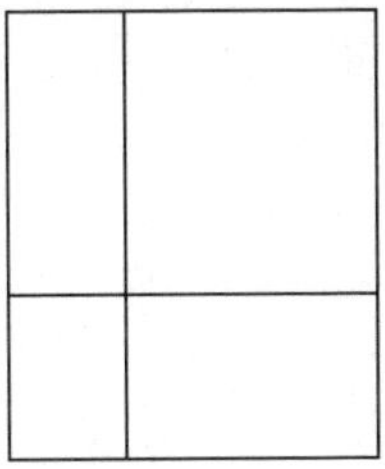

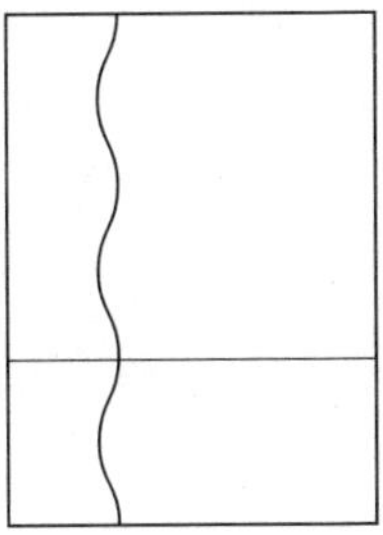

Your discussion of the possible and actual interests me a great deal. Yes, we can agree that "possibility is not *just* a subjective judgement." But I would do so in a context you would disagree with. That is, possibility is only inherent in the interaction between a human being and nature, and not in extra-human nature as such. I mean this in the same way that the act of abstraction is not inherent in nature, but only due to the presence of humans who are able to relate themselves to nature with abstractions. Hence, the objectivity of the notion of possibility has its *base* in human action, nature only providing the field for such action. One can make a very interesting distinction between the mimetic and purely creative artist's relations to nature, where we find the term "possibilities" has two greatly different contexts, and so meanings. The mimetic artist's possibilities are *predominately* determined by his capacity to discern the already present actualities perceivable in *optical* nature. Whereas, the purely creative artist's possibilities rest not only on the capacity to discern what the mimetic artist perceives in nature. He takes *optical* perception further to center his focus on the process character of nature's structure. Thus he is able to discern *possibilities that are potential to nature but are not actual in nature*, and which would never become actualities without the human artist. The mimetic artist discovers and largely records what nature has already actualized; the non-mimetic artist is a discoverer of hitherto non-actualized possibilities, which only human nature can actualize.

From the above you can see the importance I attach to the point you bring out of non-actualized possibilities being "in some sense real." This is also important because it applies to the consideration of what "has been" as well as what "can be" or "will be." This attitude has played a very important role in my study of art history, as well as in the evaluation of my personal art history. By this means I have been able to comprehend, on the one hand, the development from Monet to Mondrian, to understand how these particular possibilities were actualized. On the other, I was able to take this goal secured by these artists, and discover possibilities they could not discern, to then formulate a more direct, more clear path to the goal in question. This becomes important for the learning process of each new generation of young artists. For this reason I never encourage the young artists who contact me to simply follow the particulars of my own development. Learning to become a creative artist is itself creative. Contrast this with the learning process of mimetic artists, I mean the truly mimetic artist who no longer exists. He not only imitates nature, but has to learn to imitate past painters and sculptors as well.

Of a given moment of time and *all* its possibilities. You say that if we knew all its relations to the totality, this would narrow that moment to what I gather you mean, its total actualized character. In the view of totality as a constant *process* of creative growth, such knowledge could only be of a moment that "has been." To know all the relations of the moment to totality, we would have to know more than you indicate. We would know not only all the possibilities whose relations have been actualized, but also all those that have not and will not be actualized except in a cosmos not yet existent. Such a view of the cosmos corresponds strongly to all previous experience, whereas an absolute, closed totality, does not do so at all. No one has been able to give a convincing picture of a beginning or an end or limit in *all* directions, to the cosmos. Even the beginning that some astrophysicists formulate, may be but another evolutionary cycle of the cosmos.

The dualisms, the oppositions of the lawless and lawful, etc., remain a mystery to me. Every time you begin to substantiate the dualistic attitude I have the experience, although this may be but my own failing, that you slip away from it into another problem. There are some silent, that is unexamined, or perhaps not presented assumptions respecting this attitude.

I know Carl Visser, but not personally. I do correspond with his friend Joost Baljeu, the editor of *Structure* whom I have known for the past four years or so. But due to some strange behavior on his part for the past several months, I have the impression our association is about to terminate. Pasmore has helped himself to my art and writing. But he is not an exponent of Structurist art, only of Pasmore.

I look forward to another good letter from you when you have the leisure. Meanwhile I wish you a good trip and stay in Israel.

My best regards
Charles Biederman

P.S. What did Einstein mean when he expressed "surprised thankfulness" that four equal rods can make a square, since, in most universes that he could imagine, there would be no such thing? This was quoted by Russell in *The New Leader*, at the time of Einstein's death.

3, Berkshire Road
Bristol 7
Glos., England
November 17, 1960

Dear Mr Biederman:

I was very glad to receive your letter of Oct. 3, which contained a great many interesting and important points. I will send you a brief (i.e., relatively brief) letter now commenting on these points, and then in a later letter, I will comment on your views on art. My reason for further delaying a discussion on art are twofold: 1) I know very little about it and require some time for my opinions to jell; 2) I have the impression that my reactions to Structurist art are closely bound up with the questions that we are discussing, so that I could clarify these reactions by clarifying these questions. In addition, I have my own reasons for discussing these questions, as they are important in the work I am now doing; and I feel that to discuss them with someone, such as yourself, in the different but related field of art, may bring out something new.

(It is like the idea of "conversation" rather than "communication" – something new may arise between us.)

Perhaps the main difficulty in the question of opposition of various categories such as necessity, contingency, etc., comes from the tendency to regard both opposites as separately existing entities, which struggle with each other and eventually unite in a happy union. Let us instead consider opposition from the point of view of *process*. Every process has one or more *pairs of directions*. These pairs of directions can be understood as movement between two contrary states; i.e., states which mutually exclude each other (so that the existence of one implies the non-existence of the other).

The contrary states are not themselves generally existent, but they are simply the *limits* of the process, which would be approached if the process continued indefinitely *in one direction*. Now every process moves *simultaneously* in both directions. The opposition is not between the limits of the process, but rather *in the two directions themselves* (i.e., oppositions in movement).

Let us look into this question more deeply. Two basic concepts in the conceptual understanding of the world are universal and particular. Something universal refers to the whole, the totality (including itself if it is truly universal). Something particular refers to some part of the totality, i.e., to less than the totality. Now, *every* finite thing that we consider is particular. Hence, particularity applies to the totality, so that *particularity is universal*. Similarly, universality is a particular, since universality is not particularity, and therefore universality does not include the totality of all possible concepts (which must include particularity).

We see then that these two opposing concepts are very closely bound together by the process of logical reasoning. In order to understand the relation better, let me draw the diagram below. I consider, for the sake of argument, two times, t_A, and t_B, and a series of points, A_1, A_2, . . . A_n, B_1, B_2, . . . B_n, which are present at times t_A and t_B respectively.

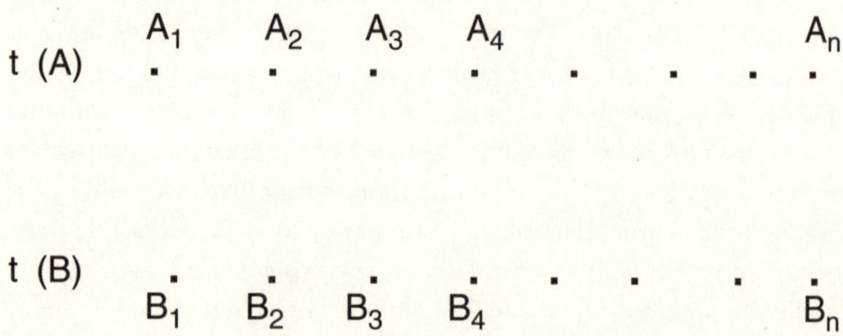

Now I consider a *process* from t_B to t_A. I say that each point $B_1, \ldots B_n$ splits up and goes to make up the points $A_1, \ldots A_n$, as indicated by the lines from t_B to t_A. I show a typical case below, where I show how B_3 splits up (you can imagine rays going out from B_3 to $A_1 \ldots A_n$).

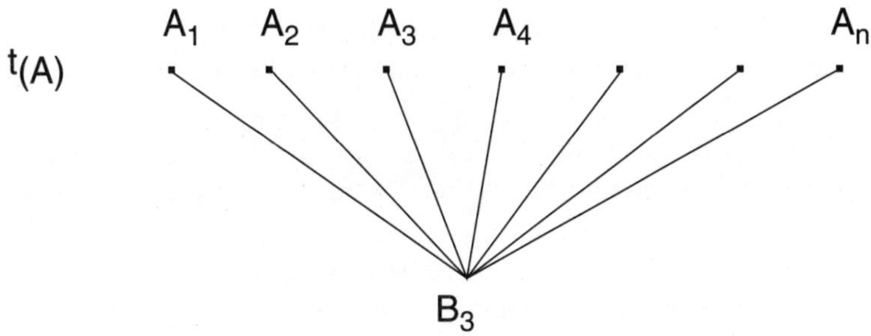

Now let us consider that all the other points, $B_1 \ldots B_n$ split up in the same way. The total effect appears in the diagram below.

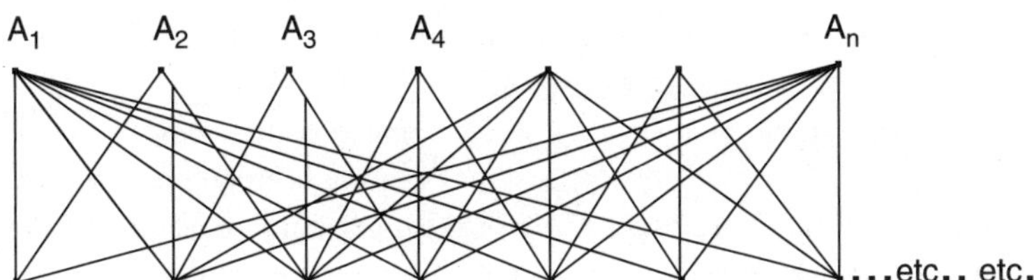

From every point B there is a set of lines connecting it with every point A.

But now there are two ways to consider these connections.

1 A point, say, B_n, splits up and enters into *all* the points A_n. Let us regard the totality of points A_n as the universe at the time t_B. Then if we start from a given point, B_n, we obtain a process of *universalization*, i.e., some particular point, B_n, becomes universal.
2 We can consider the process from the opposite side. Each point A_n is made up of contributions from the totality of points B_n. This is a process by which the universal (the universe at the time t_B) divides itself and forms particular points $A_1 \ldots A_n$ at the time t_A. It is therefore a process of *particularization*, a process in which the universe becomes particular.

The two processes of universalization and particularization are by definition opposed to each other in direction. Nevertheless, in the totality, they are identical. For the totality of the process of universalization (represented by the totality of all the lines from t_B to t_A) is exactly the same as the totality of the process of particularization. Thus, we have in the *totality* an identity of opposite processes. However, in any part, we have a diversity of these opposite processes. For, considering any point, say B_n, we have a process of universalization away from it, and considering any point, say A_n, we have a process of particularization toward it.

Here, it is important to stress that it would be a distortion to discuss the *totality alone*. For the totality is a *totality of parts*. Thus the parts are as real as the totality. Hence the identity of opposites is neither more nor less real than is their difference. The truth is that the same process opposes itself, and that the opposites are identical in the totality, different in the parts. Now, we come to a further essential point. In accordance to what I said before, the reality is the process itself, viz., universalization and particularization. The universal and the particular are only abstractions; viz., the hypothetical limits of this process in opposing directions. Therefore, in my diagrams, you should think of the points as abstractions in comparison to the process itself, as represented by the lines.

I would try to express other oppositions in a similar way. For example, consider potential and actual. These are two opposing processes: (a) the potential becomes actual; (b) the actual becomes potential (actualization and potentialization). In the totality, the two processes are identical. But in each part, one or the other will dominate. Likewise, neither the potential nor the actual exists as such in a literal sense but they are just the abstract limits of the process extrapolated indefinitely in a given direction.

Let us now try to look at necessity and contingency in a similar light. Likewise we say that necessity and contingency are only abstract limits of the process. It is a bit clumsy to find appropriate words for the two opposite sides of the process, but let us call them "necessitation" and "contingentation". In other words, there exists nothing in the universe that is absolutely necessary or absolutely contingent. Rather, there exists a process in which the contingent becomes necessary and the necessary contingent. In the totality, the two directions are identical, but in any part, one side is emphasized.

Let me try to explain this farther. When one says that something is contingent, one says that it is a part, dependent on the totality, so that the reason for itself is not in itself alone. But as we have seen, each such a contingent part is the limit of a process of particularization. *This part does not and cannot exist by itself as such.* Rather, its very being a part is only an abstraction, a limit of the process of particularization by which this part exists. But the part is also always in a process of universalization, a movement by which it falls back to the totality. Each part is only a moment in the above process, in which the property of being a part (i.e., "partiality") is emphasized in relation to universality.

Insofar as we consider the total process of particularization, we find in it a movement towards necessitation and towards its opposite. The movement towards necessitation can be simulated by considering more and more of the lines in the process (see my diagram). For example, if we see several lines moving towards a certain point, we conclude that there must be still more lines aiming towards intersection at that point, and that it cannot be otherwise because of the point character of the pattern. You might object here that the movement of necessitation occurs only in our own minds. However, I wish to assert that the movement in our own minds is a reflection of a movement that occurs in reality; viz., that with the passage of time, more and more lines come into existence. Thus, there is a process of necessitation in reality, reflected by a corresponding logical process in our thinking.

What is the opposite of necessitation? This would correspond to a falling away of lines (not only in our thinking but in reality as well). In other words, some entity (e.g., a point) becomes necessitated as it establishes more and more links with the totality, which makes its change more and more difficult. (Necessitation may be equated with causation, provided that we think of causation as an *active process* by which something is maintained in existence and not just a correlation of past and future.) But then an existent entity may lose its links with the totality, thus becoming less necessary; and eventually, it may then change or even pass out of existence. Thus, something which is, at a given stage of the process, mainly a contingency will become more necessary later, and eventually it may fall back into being more of a contingency.

My view is then that necessitation is a process; viz., the establishment of links with the totality, which tends to fix something in the state in which it is, while contingentation is the opposite process.

To sum up my view on duality and process. I say that every process is in itself dual, since it has opposite directions. In the totality the duals are one, and in the partiality, they are different. Both the unity and the duality of the dual sides of the process are equally real and equally essential. What are usually called opposites are abstractions, the limits of each direction of the process. We may therefore call this point of view the unity in duality and duality in unity, this being the meaning of a process. That is to say, the very idea of a process contains this conception, since a process cannot occur unless there is a change and change cannot occur without a duality of directions.

I think there is a close relation between the above and some things you said in your last letter. You agree (p. 2) that "when creative abstractions are made before a future as it comes into experience, the necessity . . . *is not* the particular abstractions made". Rather, as you say, necessity lies in the fact that the abstractions will be like the last ones, in the general way in which they are related to the determinant structure of nature and of human life. Therefore, it seems to me that in effect, you admit that *to begin with*, the particular has an aspect of contingency. In other words, the past structure of nature and human nature does not determine each individual creative abstraction in all of its detail. This aspect of contingency is objective, and not just the result of our way of looking at things. For as we both admit, each moment really has its past (i.e., the existence of a past is not just due to our inability to know the future, it represents a genuine completeness in the structure of the moment itself). And in relation to *this* past, the next creative abstraction is in certain ways contingent.

Where then does necessity come in? The answer is that if a particular creative abstraction is fruitful, viable, correct, or whatever else you wish to call it, it will be able to establish more links with the totality of natural and human reality. In this way, it is *necessitated.* Abstractions that are arbitrary will not do this, so that they will quickly change or die. *Nothing makes their existence necessary*. It seems to me that this is the problem of all artists, and especially the Structurists as it is explained in your book, articles, and letters. For example, as you say on the last page of your letter (p. 6), you study the development of artists from Monet to Mondrian, to understand how certain possibilities were actualized, and from this to see further possibilities that they did not discern. As I see it, this means that each of these artists was able to propose something that was for some period of its development *necessitated.* Then as the development proceeded, it began to lose its necessity, and to peter out. You are trying to see how to continue the development, so that it establishes further links to the totality of reality, in order that the process of necessitation shall go on.

The above shows that each new creative act, while initially contingent, must prove its necessity by fitting into the totality of human and natural reality. Those things which are not necessitated again and again in this way will die. Thus, as Ibsen once said, "Truths can grow old" (unless they develop in response to the new total situation in which they exist, *a situation to which they themselves contribute an essential aspect*).

Now to come to another point, namely, your notion that the concept of possibility applies only to the relation between human beings and nature. (For example, you say there would not be contingency if there were no human beings to experience it. Also on p. 5, you say that possibility is inherent only in the interaction between a human being and nature.) I want to give some arguments in favour of the point of view that possibility is a universal category, existing both in nature, in human beings, and in their relation. In order to do this, I want first to explain in more detail my ideas on totality, as it seems I didn't make them clear enough until now. I want to say that I agree with you in that there is no absolute fixed and final totality. Rather, I would say that each moment, each "here and now" is an infinite, inexhaustible totality. It is infinite and inexhaustible because this moment must be related to the whole of *its* past, and to the whole universe through this. I gave an image of this relation in a previous letter, in terms of waves converging from infinity to each point in space and time and then falling back into infinity. Each point is therefore the product of a process, to which the whole universe and the whole of its past contributes. For this reason, each point will *be* a kind of reflection of this past.

Let me now give a more general argument. When we measure space, we usually imagine taking a ruler or a tape from one point to another, and counting the number of divisions in the line that joins the two points. In the measurement of time, however, the situation is very different. We cannot take a clock and go back to yesterday, in order to measure the time between today and yesterday. Rather, we must have present *now* (today) a trace of yesterday. A trace may be a memory, a photograph, a record made by a man, but it may also be a natural trace (e.g., tracks of a dinosaur or layers of folded rocks). A trace is something that is *not* the past, but which is a *function* of the past, a structure in which the past is, as it were, folded and tangled up, but in principle capable of being unfolded and untangled, at least in our minds. Even more, it may be said that one side of what exists at each moment *is* just this tangled and folded trace of the past. This applies not only to each time, but to each place. Each moment, as it exists, is a "here and now", containing its past as trace. But if, in principle at least, we are to be capable, at each moment, of knowing the whole past, it follows that this must be continued as a trace in each "here and now". For if there were a "here and now" that did not have the whole past (and indeed the whole universe) folded into itself in the form of a trace, then there would be a place and time for which some of the past (and some part of the universe) would in effect not exist. Since there is no such place and time, it follows that each part of space and time has the totality of its past and the whole universe folded into it. This applies not so much to nature as to man. What I am trying to say is that the world is *as if* every moment were *potentially* conscious of the universe, in the sense, that in its inner structure, the whole history of the latter would in principle be read. In this regard, the moments experienced in man's consciousness could be on the same footing as the moments of existence in nature. Man is part of the totality, which includes himself and nature; and in the general character of space and time, he shares these properties, which are universal.

Let us now proceed further. Given any moment, A, it contains a folded trace of previous moments (or is such a folded trace). Now consider a later moment, B. This contains a folded trace of A, as well as of all moments which are contained in A. In this way, we see that B is *inwardly and in itself* later than A. For whereas A is folded in the part of B representing the past, B does not similarly appear in A. (It is *as if* B remembers A while A does not remember B – but memory in the sense that I use it is universal.)

It is clear then that a given moment (say B) not only contains (and is) a *totality* of the traces of previous moments, but that it is also a *totality of totalities.* On the other hand, B itself enters into still later moments (such as C) as part of their folded structure. And so on. It follows then that no totality is complete. For each moment has a *future* which will be determinate only in the past of still later moments.

In this regard, we must see this process dually. The folding of the past into A is the process of particularization, which I indicated diagrammatically earlier in this letter. It is through this process that the past of A is present – i.e., "presented" to A. Thus, presence is a relationship (but an inner relationship). There is no such a thing as the universal present, but rather there is the present of each here and now. We must remember, however, that each here and now is also thus present to all the moments in the future. This is the process of *universalization*, whereby A breaks up; as it were, and re-appears, folded up with other events, in the past side of all later moments. As I explained before, in relation to the totality, the two processes are the same, while in relation to each "here and now" they are different.

Naturally, in any moment, there cannot exist a complete and fixed totality. To discuss the totality of a given moment, we must go to the limit of the infinite future of that moment. Then its totality will appear on the past side of this infinitely later moment. It is evident then that totality is (like partiality) an abstraction, which does not exist by itself. What is immediate is the partial moment, the individual "here and now". But what this moment *is* in its totality is only revealed in the totality of later moments. Since the part and the totality are equally necessary for the process to be what it is, it follows that the unity of opposites in the totality and their diversity in the parts are on an equally valid footing.

It is interesting to point out that the two sides of the process also represent *passivity* and *activity*. Thus, as the past of a given moment appears folded up as trace, the moment plays a passive role, being formed by its past. But then, as that moment breaks up and re-appears folded up in its future moments, it is playing an active role, entering into and helping to form its future.

Now let us return to possibility and actuality. As it exists, each "here and now" is actual, and in its future is a range of possibilities as determined by its past. In the next moment, some of these possibilities have been actualized, and new possibilities are created. The question of what will be actualized depends on the interaction of this moment with all the others. Notice that I am giving to the "here and now moments" existing in nature a role similar to that which you give to those existing in man. I say that nature is in itself creative. Man is creative too. While his mode of creativity is peculiar to himself, his being creative is a special case of a universal process of creativity. As a result, possibility has meaning in the relation of any "here and now" to the totality; viz., that given what exists in the past of such "here and now", there is a range of possibilities open to developments in the future of this here and now. This range will be narrowed down when "more and more here and now's" are taken into account, as when what is in this moment comes to interact with more and more similar moments,

Finally, it is clear that in this point of view, contingency has an objective meaning. For each "here and now" is as essential and real as any totality. Given the past of any "here and now", say A, the next moment, B, has aspects that are not necessitated by anything in the past of A. Of course, in the total process, which includes B, A, and everything else, it will become clear how any feature will cease to be necessary in a later moment, C, so that this feature will not continue to exist. Thus, there is objective contingency in relation to certain moments, and objective necessity in relation to other moments, as well as a process of necessitation in the passage from one moment to another (and also of course the dual process of contingentation). In a sense, we can say that judged from the standpoint of a certain moment, something can be contingent. One of the moments under discussion may be the moments of consciousness of some human beings. Events may therefore be contingent in relation to him, too, but this is only a special case of the general relation of contingency,

With regard to possibility, it is the same. Which possibility will become actual depends on the totality, including the human being. A human being may look at nature and see possibilities, which can be determined only by the totality, including himself. But if we consider any moment in nature, we can say that the *rest of the world* has certain possibilities, which are fully determined only in its relation to *that moment.* Therefore, when the human being creatively determines certain possibilities to be actualized, he is participating in a universal process.

I have been using the words, unity and diversity of opposites in certain relationships. To be more consistent, I should perhaps have referred to the process of unification and diversification. Complete unity and diversity are only the abstract limits of this process. Thus, as we consider more and more moments, towards the future, of a given moment A, we approach the totality for A. We never reach it. Totality is therefore the limit of a process. Similarly, absolute partiality, particularity (the mathematical point) is the limit of this process in the opposite direction. But I claim that this process occurs not only in our own minds, but also in nature. In other words, each "point" breaks up and spreads out, tending to universalize and "totalize" itself, while the dual process takes place, by which the contributions of many points move towards unity as "one-ness".

As usual this letter has gotten longer than I expected. However, I think that after these questions are clarified, we can get down to discuss artistic problems in a fruitful way (at least I hope so). I hope that I have already indicated the relevance of some of these questions to Structurist art.

Now to answer some of the specific questions that you asked me. On p. 5, you give an example of a geometrical form with two lines, and discussed the multitude of contingencies that have to be removed.

The example is perhaps too simple for me to give an adequate answer. In terms of mimetic art, I can more easily express what I wanted to say. Thus, if we have the picture of a man, then the spots of paint representing his eyes are in *themselves* contingent in that they could be located anywhere in the picture. This contingency of location is removed when we consider the whole picture, which is supposed to represent a man.

Now a picture is static, so that in its structure, it cannot express contingency and necessity as a process. The process must be in the mind of the man who views it. When he first sees a given item, it is "just there", i.e., it is contingent, then he sees more, and gradually, the picture as a whole is registered. Then he begins to understand (perhaps by a creative act of his own) the "law of a picture" which is its essential content, and which begins to necessitate each detail *if it is a good picture.* If it is not a good picture, he may discover that the contingency of some parts is only reinforced by his overall view of the picture. Or he may experience necessitation of a given part in a cursory viewing, but see its contingency on a more careful viewing. So you see, the viewer passes through the process of necessitation and contingentation. Presumably the artist does too. The creative step is the realization of the concept of the picture as a whole, but then he must execute it in paint and see if it remains necessary. The further development of his own and other people's painting may also either necessitate or de-necessitate the line that he has been following, as well as the development of techniques, society, etc.

With regard to Structurist art, I imagine that the problem is similar, except that the overall law in the picture is not determined by the mimesis of an object, but rather by the mimesis of the general laws of structure in nature. Thus, the contingency of each part would be removed by the fact that it was necessary in view of the overall pattern and purpose of the work. The Structurist might even, as you say, discover possibilities that are potential to nature, but are actualized only with the aid of man (e.g., by the artist himself). But again, the removal of contingency of any part means that it is this overall content of the work which necessitates that part.

Here, I would like to stress again that all art has as its object not so much the embodiment of the processes of nature as the initiation of these processes in the viewer. As we both agree, art does not communicate, but rather, it helps to guide a creative process in the viewer. Structurist art should, according to this notion, set up in the viewer a process which reveals, not a particular object, but rather, the general laws and possibilities of nature, the essential structure of reality. If it does this, then it will itself be a reality, and not just a copy of something else. For the essential character of reality is that it is constituted by such processes. Would you say this is a correct interpretation of your point of view?

Finally, with regard to your post-script question about Einstein's statement expressing "surprised thankfulness that four equal rods can make a square, since in most universes that he could imagine, there would be no such thing". Here Einstein was referring to certain aspects of non-Euclidean geometry. In particular, he supposed that he started with four equal rods, their equality being demonstrated by the fact that when they are placed side by side, their end points coincide. Now, to make up a square out of these rods, it is necessary to displace three of them in relation to a fourth. Let us do this in steps. Let the fixed rod be A. Then let us displace another rod, B, parallel to itself through a distance equal to one of the rods, as shown in the first diagram.

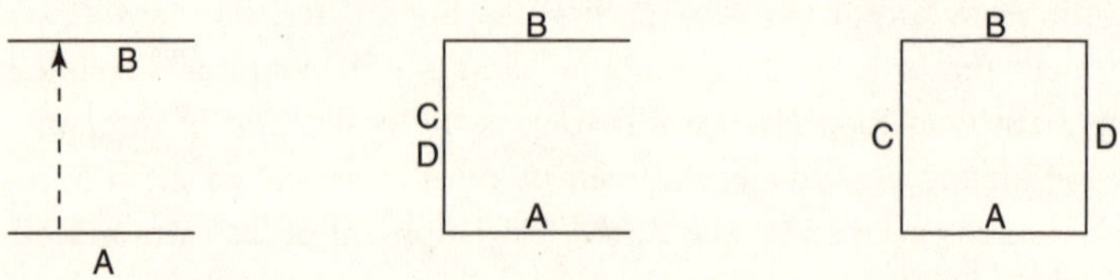

Next we take rods C and D and rotate them 90°, forming the second diagram. Finally, we displace D parallel to itself to form the square.

Now, in non-Euclidean geometry, one goes more deeply into the question of what is meant by parallelism and by a right angle. In a more general geometry, it is possible that initially equal rods displaced parallel to themselves, in the way described above, will in effect cease to have equal length. Thus, the square would not close, as I indicated in the diagram below.

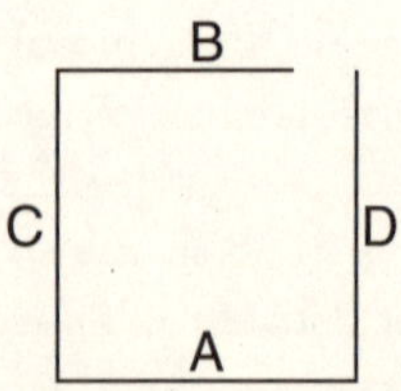

We cannot imagine this behaviour in terms of the space of ordinary experience. Yet logically, when we study the postulates underlying geometry, we can see no inherent reason why this latter behaviour (the non-closing) should not be possible. The fact that they *do close* is then empirical, rather than logically necessary. This means of course that the empirically observed closing may only be an approximation, and that eventually a very accurate measurement would show otherwise. Similarly, perhaps the closing is characteristic of our ordinary scale of distances, and might not apply either on the sub-atomic scale of smallness or the super-galactic scale of largeness.

I hope that this gives you some idea of what is involved here. It is very hard to give a more precise idea without mathematics.

With regard to your request for books about this view on creativity in nature, it is hard to suggest anything, as few people hold these views at present. I would suggest your reading Whitehead as the nearest approximation that I know of. For example, try his book, *Process and Reality*, Cambridge Univ. Press, 1929.

With best regards, and looking forward to another letter from you,
David Bohm

Red Wing
Route 2
Minnesota, USA
December 28, 1960

Dear Mr Bohm:

Your letter of November 17 was read with interest in what you have to say. However, my reply is handicapped by the fact that I could not understand several of your key points.

You must forgive me but I had to smile at your explanation for delaying discussion of art. For everything you say applies as much to me as an artist trying to discuss your work. Nevertheless, since I have taken the plunge first, and I assure you it is a plunge, it is just as well you delay your remarks on art until our conversations give you something to go on. I am hoping that we will also unearth those psychological differences and relations between physics and art.

For instance, I have been struck for a long time by the necessity for the physicist to minutely ponder whatever he encounters before that aspect of nature which concerns him. As an artist I never feel the need to pursue my thoughts to the nth degree, although I do so more than most artists, as do those few who are concerned with the new direction. If I were to pursue art as the physicist pursues his, much as I would find it pleasurable, I would be making inroads into my main work, to be constantly experiencing and seeking understanding of vision for art purposes. This reminds me, several times in the past I have read that scientists are generally more interested in music (sound); do you find it so?

It is interesting to note, that the field of philosophy called aesthetics has tried to formulate a minute penetration of art experience. But it invariably turns out to be quite useless, ignored by artists, because so far removed from the actualities of art. As has been noted by others, it could go on existing even if there were no art.

You start off by saying that every process has one or more pairs of directions. How do you adjust this view with the often stated notion that the general cosmic process is a one-way, irreversible affair? I think I understand what you are talking about as "contrary states," these "oppositions in movement." You say that these states are not "generally existent," but it seems to me they are never existent. There are only the interweavings of different components of a process, in which none can achieve complete supremacy to its own limits.

I don't see how these states, really aspects of a single process, are seen to possess directions "contrary" to, and in "opposition" to, each other. Incidentally, are you familiar with Piet Mondrian's Neoplastic theory of art? In it he has his "oppositions," dualisms and universals.

Let's take the problem of symmetry which I take to be characteristic of all processes in nature and human life. In your language, if one extended the two aspects of the process to their limits there would be absolute, frigid symmetry in one direction, asymmetric chaos in the other. In actuality we never face either extreme. What we commonly call symmetry is never without its asymmetric aspects, the latter never without symmetric aspects. Thus the cosmos does not freeze into rigor mortis, nor disintegrate into utter chaos. To make contraries out of that is to have the choice between hanging and shooting. What I do find is a *spectrum* of variables disclosing the rich order that is extant in the creative structure of nature's unity. Both aspects are always present together in varying degree. This corresponds to what you call their "identical" characteristic in the totality. But I question the term identical, preferring your reference to the degree of dominance of one aspect or another. So I don't find any opposites which are even opposite to themselves (because they are said to achieve identity too), where processes themselves are opposite to themselves. You start with contraries and end up with contradictions.

What you say about universals escapes me. The following sentence completely baffled me: "Similarly, universality is particular, since universality is not particularity, and therefore universality does not include the totality of all possible concepts (which must include particularity)."

Does necessitation occur only in our own minds (either–or)? Necessitation or determination of some kind will probably continue to be regarded as an attribute of nature in the future. But in that future aspects of our present view of necessitation will be found to lack existence outside our minds. In this qualified manner I would admit that our mind reflects what occurs in reality. If it did not, we would not have a reliable means for discerning error, and if the reflection were perfect, however, we would never make errors only experiencing perfect reflections of each new moment. It appears, then, what we experience outside us is derived from what is neither words, thoughts nor perceptions. These latter are attributes peculiar to the human aspect of nature and remain as such, no matter how well the structure implied by our words, thoughts and perceptions appears to correspond to, even to reflect nature. The reflections are *limited* reflections because the capacity of our mind for reflection is, at any one time, always *limited*. Thus our mode of perception and/or abstraction when *projected* upon nature seems to achieve a correspondence of structure within the nature context being experienced. Our projections seek structural consciousness that corresponds to a nature that *appears* to be reflected. The history of art offers vivid examples. Until recent times when some have become more acutely aware of the non-allness of our perceptive and abstractional abilities, verbal statements were made along with contemporaneous art, especially from Giotto to Courbet, exclaiming that nature's realism had been captured to the nth degree. The reflections of nature appeared to correspond to the art, and appeared so because nature was not perceived, not consciously reflected, to a degree greater than that reflected in the art. Therefore the verbal statements of realism only indicated the conscious condition of human vision (reflection) of nature at the time.

The notion of duality, you say, is essential if we are to be able to show the interweavings of the different aspects of a process, otherwise we could only make the trivial statement that the cosmic process was One. It is not my intention to deny the different aspects of any process; only the inference that these aspects are in "opposition." To return to our discussion of past and future, I suppose you would say there were two directions, a going into the past of each moment, and each moment going away from the past into the future. Are these oppositions, or opposite directions? Simultaneously, each moment becomes past and future at the threshold of the present. What you have so well explained as the folding of each moment into the past, is also a folding into the future. There is, however, only one direction to the process, the future. Strictly speaking, there is no past, "What has gone has gone forever," as Whitehead put it. As you say, we cannot go back to yesterday with a clock and measure the time to today. There are only various "traces " of the past "folded" up in the present "here and now." The traces themselves are constantly moving, in motion, *from* what they were towards an inevitable series of will be. Traces of traces are left behind. Truly, everything that was has indeed vanished forever. What remains of the past is what it was not. With these qualifications I can agree the past continues to exist, the term "continues" implying the qualifications anyway. Only our ingenuity to abstract permits us to unfold past times, places and things and even thoughts and perceptions of the past. I do not find the passive role you refer to, only a past that is actively responding to the alterations of the future, as the future is constantly responding to the residue of the whole past which, as you say, is contained in each whole moment.

In your remarks on arbitrary abstractions, you couple them with mine on the development from Monet to Mondrian, as an example of what was once necessity then "petered out." On that quoted remark we may not be in agreement. True, I study the development from M. to M. in order to comprehend the specific structure of the transition from mimeticism to creation. This leads to discovering certain possibilities that were actualized by these artists which made possible and sustained this transition, as distinct from other actualizations which were in direct conflict with their goal. It is these latter qualities, some extant some suggestively potential, in the art of Monet, for instance, which led to the arbitrary abstractions of all Monet's contemporaries, except Cézanne. It is just these arbitrary efforts that have been leading the majority efforts at art steadily to a death that is now close at hand in today's art. There we see that art has become a mere therapy for the artist himself, but a reverse sort, in that they only release the urge to destroy. This led me to divide all post-Courbet art into Structurism and Expressionism, in my Cézanne book. Here indeed I certainly see oppositions, dualisms, contraries. *What you want to attribute to nature, I attribute exclusively to man.*

The non-arbitrary actualizations remain sustained because they remain necessitations. As we increase our understanding of these necessitations, they continue to play a part in clarifying the particular necessitations of future art. Thus, anything that once attains true necessitation in art, remains so, for we have yet to exhaust our understanding of the very first art attempts of cave artists when they began picturing the world in their eyes. As you said with Ibsen, the truth can grow old, but it can also respond to the new as it comes.

Such problems, however, are not fully recognized in art. The belief is that the artist is free to do as he pleases. The concern is with whatever the artist does, not whether it is desirable. In this way a Picasso can be quite sincere in announcing the demise of Cubism, without knowing that it was his false understanding of Cubism that led to its demise. He cannot know that Cubism is open to a different view, that it can be actualized otherwise, and in a way in which its necessitation is sustained.

Contingency is peculiar to human life. You disagree. Perhaps my position is between the one you take and the one you think I take. Nothing in experience gives me reason to regard the objective reality as a process containing contingentation, to believe that anything could be other than it is at a particular moment. I do believe, however, that the unique structure of human life allows it a multiple of potentiality choices before each future, that things could certainly be other than they are. The history of the human race is full of examples of this multiplicity. And today, in both the fields of science and art, we see similar material acted upon as having different potentialities. This leads me to believe that in some respects nature's objective reality could be otherwise where the intercession of man is possible. This is to exclude a belief in chance.

If I stubbornly reject chance as an inherent aspect of nature's structure, it is not that I count on the hope that were we given knowledge of the infinite future, causality, as we know it today, would hold firm. Looking at all man's past searchings into nature structure, we see his vision of it change too often to have any hope (or desire) that current views will last any longer than so long, and then be replaced by others. What I feel, I don't mind admitting that it's instinctual, is a notion of causality remarkably altered by a nature seen as a purely creative process – not the atomic particles, but what is at the bottom of that structure, not the objects we perceive with our eyes, but what is at the bottom of that perception. A creative causality that rejects any form of mechanistic thinking equally with any notions of chance. A process where what now appears as chance is but the subtle intricacies of creation, causality as a process of diverse orders of creative determination.

Chance theories of nature, and here too I don't mind admitting I have only my feelings to speak from, to the degree held seem to me as signs of man's defeat before nature. Men claim that a stone wall exists at a certain point in his penetration of nature, which he proclaims both absolute and impenetrable, thus assuring himself that there is no need to keep on looking, on the *chance* that he, not nature, erected that wall. Had physicists taken art more seriously than they generally do, they might have detected a warning to themselves at the sight of many in both art and music turning, without stopping for a decent amount of thought, to the views of chance and indeterminism as the very essence of nature and human life.

A nature with chance is just as repugnant as one of mechanistic determinism. The latter eliminates the essential quality of creation, the former reduces it to the absurd, to that creative irrationality seemingly implied by Bohr's "irrationality trait," p. 91 your book.

All my feelings, thoughts experienced with nature, as well as my attempts to follow various physicists as they recount the motions in the intimate structure of nature, all come to a nature of creative motions, growth. I feel the elements of wonder, discovery, the constant creative renewal of nature pouring into the veins of human life, to make man a unique creator within nature's process. I think early men were on the right track. Their instinctual yearnings as creatures of nature led them to form their greatest gods as beings of creation; creators ordering the creation of the universe.

In speaking this way, I am not under the impression that I prove anything, that I am talking in a way that seems to utter trivialities, lacking what is needed for their serious consideration. So why talk like this at all? Because I believe that human feelings are the place where an ever greater degree of objectivity can be born. I think this has been true of all the great art and great scientific discoveries. Our feelings are our life-link with nature, as they are with other human beings. In this way I have found the direction that I believe leads to a genuine art of our times; in the same manner, I find the way to the development of that art. We live, but to fully live we must feel. And whatever we say about the structure of nature and human life must lead us to feel it, and to always seek to feel it deeper. Otherwise, it is like making love to an apparition.

I have the impression that twentieth-century urbanism coupled with the immense power that man now wields thanks to science, the age of man-power instead of horse-power, that all this has dulled our confidence to feel. We have Geiger counters for the atoms, but we have them for humans too. These days one is supposed only to "communicate." We want to rise above feelings, because we fear to feel. It has left a vacuum in life which we leave the totalitarians to fill, they do make people "feel."

From this viewpoint, indeterminacy appears to me to have its major appeal to the intellect, to the Ego of man frustrated before a new and illusive reality appearance of nature, both in art and science.

Since we agree there cannot be an absolute fixed totality, how can any moment, any "here and now," be "an infinite, inexhaustible totality"? This is not achieved, as you propose, by relating the moment to its whole past, and the whole universe. That would be to relate it only to a limited totality, a particular of the infinite creative process of nature. Only by its relation to a *future* that extends to infinity, is any past moment infinite. You yourself, further on in your letter, show how no totality is complete, when you discuss moment A folded as a trace in moment B, etc. Still further on you say that totality is found in the limit of the infinite future of the particular moment.

I thought your exposition of how each moment was a kind of reflection of the past excellent, as were the verbal expressions such as traces, folded and unfolded. What you say explains very well my reasons for being necessarily concerned with the historical process of art as being applicable to present decisions about art. When you publish the views you are discussing with me, I will ask my more serious readers to study your work.

Regarding contingency as you demonstrate it in a painting. First I want you to know that I readily agree to contingency in works of art. Still a very interesting thing happened when I read your views on this. In the way you tried to explain contingency, it seems to gain in significance when applied to non-mimetic art. In the painting of a man your contingency remarks only apply to whether or not the parts, the eyes, etc., were properly related according to the *determination imposed by created models in nature.* This deals only with the facility of the artist to imitate, not with his creative ability. So let's expand the context. Again a man, a painting by Rembrandt, depicted in head gear and ornate dress, looking at a bust on which one of his hands rests. There is the table, etc. Here your contingency remarks really get to work. For we are now concerned with this artist's remarkable ability to creatively compose the picture toward an expressive goal.

But it is precisely this latter effort that takes on an even fuller existence where art is wholly created by man, that is, he is not limited by the finite creative area left open by the demands of mimeticism of already created objects by nature, carpenters, etc. I chose the Mondrian-like painting, limited to the context of two lines in right-angle relations, in order to avoid unnecessary visual complications requiring

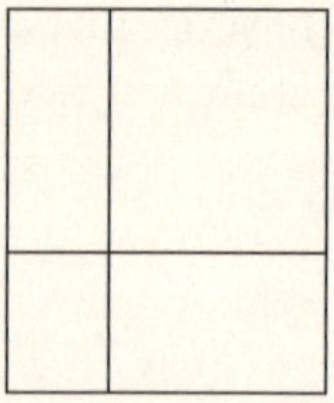

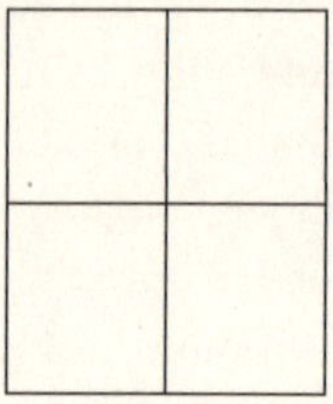

much explanation. In such a painting, and I do not need to depart from the words you used for your man painting, the right-angle lines are "contingent in that they could be located anywhere in the picture." For example, the area of the canvas could be an equal sided square, the two lines dividing it into four equal parts. "This contingency is removed when we consider the whole picture," for we find it to be a more creative expression of the lines and colors within the area. There is then the difference between a dull monotony and the creative expression of diversity, as in the case of sounds produced by an instrumental composer of music.

But there is more to contingency in purely creative art. No genuine artist would ever claim that any of his works were perfect, i.e., that all contingency had been removed, that necessitation reigned perfectly supreme. (The nearest anyone comes to claiming this perfection, interestingly enough, are idealists who claim knowledge of some absolute, as did Mondrian.) Cézanne is a vivid example who has been misunderstood on just this problem, because of his repeated expressions of wishing to "realize." The artist strives with the hope that certain contingentations in past work will find their resolution in necessitations achieved in future work, but never with the expectation (or the desire) that this process between the two aspects will ever cease. In regard to this problem, artists like Mozart and Monet were highly sensitized vehicles for their art, but men like Cézanne, Beethoven, Poussin were also seeking to be masters of the process of their art. Both kinds of artists have their *necessary* and *irreplaceable* place in the evolution of art.

Going further still, let us see how this non-absolute, non-perfection, this non-idealistic attitude, brings up another aspect of contingency. If a work of art is such that it "begins to necessitate each detail," it is then static, frozen instead of living. Examples of this art are most completely achieved by artists who use mathematics as the determinant for the structure of the parts of their art. At the other extreme we find contingency floundering across the work hopelessly searching for necessitation. Examples of this art cover the galleries of the world today, called Action Painting here, Tachiste Painting over there. It is this type of artist who are highly pleased to hear the theory of indeterminism murmured in their ears backed up by the high source of "science."

Contingency in great works of art serves the role of securing the living quality which, in the two examples above, is either chaotic or frigid. Where this living quality is present, the work seems inexhaustible to experience, we can come back to it again and again. New contingencies keep on appearing which lead us to its broader base of necessitation. Thus the totality of the work is constantly changing and expanding, until we begin to experience a fuller sense of correspondence with the art. In this way the great work of art approaches the living quality experienced by the creations of nature, where nothing remains a fixed totality. It is then not enough to merely attain structural integration in the "overall pattern," to use your phrase, for that can be achieved easily enough by submitting to the dictates of mathematics. It is necessary that the work's structural pattern be imbued with the dynamic expression of the living quality most exemplified in the creations of nature. This cannot be achieved at the price of "disorder" or indeterminacy, as the drip and splash artists think.

The structure of a picture is static, you say, so process experiences, contingency–necessity, occur only in the human being. Would you reason in the same way about a sea-shell? You seem to assume that unless a work possesses actual motion it remains for the viewer to set the static parts in motion, at least in himself. This is to neglect the structure of the eye with its focal area, to forget that the artist arranged the parts that the eye might experience the motion of the focal points coming together in certain ways. If process experience could only be manifested in visible motion of what the eye sees, the reality would be a rock-and-roll one. Nothing is truly static anyway.

Every new work of art has some contingency relations, and so necessity relations, with all past great art, and the reverse is also true. You are looking at a Rembrandt painting, the Rembrandt you see has never been seen before, not only because you have never been before, but also your vision is colored by all the art since Rembrandt, as well as before. This accounts for the rise and fall of certain works or artists with each new generation. What happens to the "static" work of art is not unlike what we have discussed in regard to past and future. It is possible to project oneself into a view that approximately sustains the contemporary experience of some past work of art. This is rarely done, it is difficult to do, yet essential for the adequate understanding of art as an evolutionary process.

"The Structurist," you write, "might even, as you say, discover possibilities that are potential to nature, but are actualized only with the aid of man," the artist. I could not have used the word "might," more likely the word "must." The Structurist takes the one alternative to mimesis – pure creation of art. He *must* realize *creations inherent* in nature's creative process, which can only be actualized by man. But even this is not too difficult to do, and that is not very well understood. What is truly difficult is to realize the living quality I referred to earlier, else the art is still-born. To the degree the particularity of the artist's form relations fail to be other than the particular relations created by nature, to that degree the Structurist fails.

I assume you mean that *all* art has the purpose of "helping to guide a creative process in the viewer." Mimetic art has two general roles – as a vehicle of literature, or performing the task which became the province of camera artists in the nineteenth century. The first involves the viewer in the creative activity peculiar to literature when it takes the form of visual art. The other reveals and stimulates human vision to experience the creations of nature. One is thankful that this last form of art continues. It is from this last that the Structurist takes the next step, revealing those creations inherent in nature's process which man alone actualizes. His art reveals to human life the deeper creative relation he can possess with nature and so with himself in all aspects of human existence. It is through this art that I came to understand science as essentially dealing with nature as creation.

In this connection I cannot agree that Structurist art has the intention of "not revealing a particular object," but of offering the "general laws and possibilities of nature, the essential structure of reality." I do not mean to say it will not, but that this can only be done with words and theory. Art is a perceptive experience, one does not merely manifest laws and their possibilities, but rather uses them as the means to manifest the creative experience of art. These manifestations possess the same qualities as nature – form, space, color. Their purpose is not so much to "guide a creative process in the viewer," as much as to imbue him, immerse him, in the quality of human existence as creation. To convey that creation is fundamental as the supreme expression of the human quality. *Art and music, however, realize in the purest form what is essential for the adequate experience and realization of human life in general.*

I appreciate your effort to explain the Einstein remark which I see, after a point, takes me in the area of my ignorance.

I read Whitehead's *Process and Reality* some 20 years ago, I quote from it in my *Evolution*. But looking into it again, at your suggestion, I find the verbiage in it still poses hurdles for me to follow consistently what he is saying.

I hope you haven't forgotten that you were going to tell me of your meeting with Carl Visser. Are he and I the only artists in the new direction that you have come in contact with? Also, I meant to ask you in my previous letter, if you are an acquaintance of J. Berger? His replacement on *The New Statesman* needs being replaced!

Again I look forward to your reply. Meanwhile, my best regards and may I wish you and your wife

a good New Year,
Charles Biederman

III

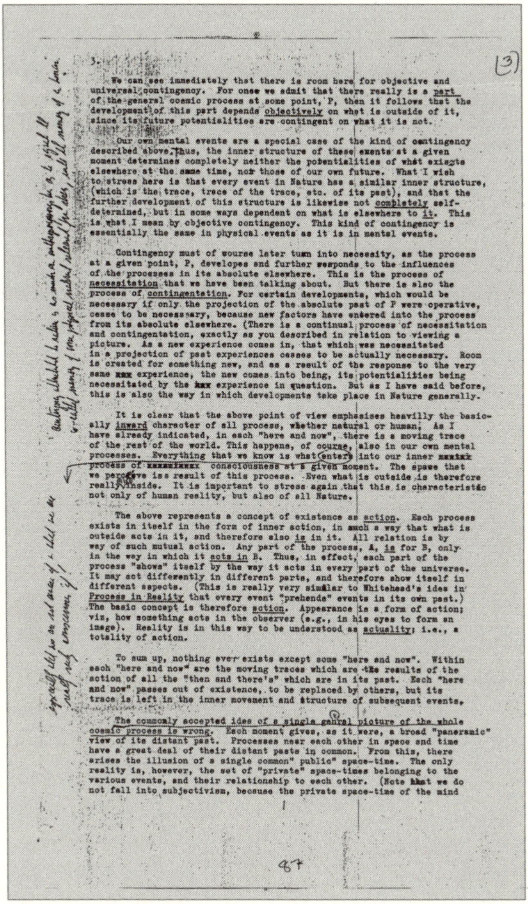

3.

We can see immediately that there is room here for objective and universal contingency. For onse we admit that there really is a <u>part</u> of the general cosmic process at some point, P, then it follows that the development of this part depends <u>objectively</u> on what is outside of it, since its future potentialities are contingent on what it is not.

Our own mental events are a special case of the kind of contingency described above. Thus, the inner structure of these events at a given moment determines completely neither the potentialities of what exists elsewhere at the same time, nor those of our own future. What I wish to stress here is that every event in Nature has a similar inner structure, (which is the trace, trace of the trace, etc. of its past), and that the further development of this structure is likewise not <u>completely</u> self-determined, but in some ways dependent on what is elsewhere to <u>it</u>. This is what I mean by objective contingency. This kind of contingency is essentially the same in physical events as it is in mental events.

Contingency must of course later turn into necessity, as the process at a given point, P, developes and further responds to the influences of the processes in its absolute elsewhere. This is the process of <u>necessitation</u> that we have been talking about. But there is also the process of <u>contingentation</u>. For certain developments, which would be necessary if only the projection of the absolute past of P were operative, cease to be necessary, because new factors have entered into the process from its absolute elsewhere. (There is a continual process of necessitation and contingentation, exactly as you described in relation to viewing a picture. As a new experience comes in, that which was necessitated in a projection of past experiences ceases to be actually necessary. Room is created for something new, and as a result of the response to the very same xxx experience, the new comes into being, its potentialities being necessitated by the xxx experience in question. But as I have said before, this is also the way in which developments take place in Nature generally.

It is clear that the above point of view emphasises heavilly the basically <u>inward</u> character of all process, whether natural or human. As I have already indicated, in each "here and now", there is a moving trace of the rest of the world. This happens, of course, also in our own mental processes. Everything that we know is what enters into our inner xxxxxx process of xxxxxxxxx consciousness at a given moment. The space that we perceive is a result of this process. Even what is outside is therefore really inside. It is important to stress again that this is characteristic not only of human reality, but also of all Nature.

The above represents a concept of existence as <u>action</u>. Each process exists in itself in the form of inner action, in such a way that what is outside acts in it, and therefore also <u>is</u> in it. All relation is by way of such mutual action. Any part of the process, A, <u>is</u> for B, only in the way in which it <u>acts in</u> B. Thus, in effect, each part of the process "shows" itself by the way it acts in every part of the universe. It may act differently in different parts, and therefore show itself in different aspects. (This is really very similar to Whitehead's idea in <u>Process in Reality</u> that every event "prehends" events in its own past.) The basic concept is therefore <u>action</u>. Appearance is a form of action; viz, how something acts in the observer (e.g., in his eyes to form an image). Reality is in this way to be understood as <u>actuality</u>; i.e., a totality of action.

To sum up, nothing ever exists except some "here and now". Within each "here and now" are the moving traces which are the results of the action of all the "then and there's" which are in its past. Each "here and now" passes out of existence, to be replaced by others, but its trace is left in the inner movement and structure of subsequent events.

<u>The commonly accepted idea of a single genral picture of the whole cosmic process is wrong.</u> Each moment gives, as it were, a broad "panoramic" view of its distant past. Processes near each other in space and time have a great deal of their distant pasts in common. From this, there arises the illusion of a single common" public" space-time. The only reality is, however, the set of "private" space-times belonging to the various events, and their relationship to each other. (Note that we do not fall into subjectivism, because the private space-time of the mind

1

87

THOUGHT AND REALITY

3, Berkshire Road
Bristol 7
Glos., England
February 2, 1961

Dear Mr Biederman:

Thank you very much for your latest letter, and the two journals, which I have read with great interest. I find that we are very close to agreement on many points, and that the respects in which we disagree are now more clearly defined. I think it best that we go further into these philosophical questions. Meanwhile, I am considering my ideas about Structurist art, which will probably be ready for discussion by the time I get your next letter. I have seen an exhibition in London of Structurist art and have had an interesting talk with Anthony Hill. When I get around to making my comments on Structurism, I shall also discuss my talks with Carl Visser and Anthony Hill. In addition, your books and articles have been lent to students at the Slade School in London, and I hope that we shall be able to have some discussion on them soon.

Your discussion of contingentation and necessitation in art is very illuminating; it expresses just what I would have liked to say. I am very surprised indeed that you do not see that the same process goes on in nature. Incidentally, may I ask you to think about a certain question related to this subject: "If man has error and contingency in the make-up of his mental processes, and if he evolved solely out of nature (and is sustained solely by his inter-connection with nature), then where did these features of his process originate, if they were not, in some sense, in nature from the very start?" In other words, do you think that man's claim to distinction is his "creation" of error and contingency out of nothing?

About the difference in emphasis in physics and art, this arises because the physicist seeks laws having some kind of universal truth. By this, one means that these laws operate in every context in which the terms mentioned in the laws exist. For this reason, a discrepancy between theory and fact, even in the smallest detail, becomes extraordinarily important. As a result, everything has eventually to be worked out in full detail. Sometimes the smallest discrepancy between theory and fact reveals the need for a revolutionary change in the whole set of concepts. As I see it, artists do not generally work in this way (except for a few like Leonardo da Vinci who was very much oriented towards science).

It is true that physicists are generally more closely attached to music than to art. Music is of course, as you say, more removed from mimesis than art, and in this sense more universal. It also has a more clearly defined relation to mathematics. What do you think?

I think that it is important to me to state my ideas on process once again, because they have evolved since the last letter was written. The main change is that I now admit (in agreement with you) that process exists even in so-called "static" objects, such as a picture. Of course, we realize that no object is really static, but that rather, its persistence in a certain form is a consequence of inner movements that balance each other (e.g., atomic movements, pulsation, etc.). But what must be added is that for this reason, the persistent aspects of the form of the thing can, and indeed must, reflect process. Your example of a sea-shell is extremely cogent. I like to give the example of the geological strata on the rocks, which reveal their process of formation, deposition, etc. These examples are only special cases of a very general principle; viz., *all structure is the trace (and the trace of the trace, etc.) of process*. But secondly, structure also *conditions* subsequent process, both in the object having the structure, and in everything else in which this object can act (including of course the human viewer). In other words, *existing structure determines potentialities for subsequent process*. Thus, in effect, process congeals as a set of persistent structures, while the real inner and outer movement of these structures implies their further unfolding and revelation of self in subsequent process.

Let us now return to the problem of time. Consider any particular moment in any particular place. This is for itself a "here and now". For every other moment, it is a "then and there". Each such moment is in contact with other moments in two ways:

1 Its past acts in it, to form a trace, trace of a trace, etc.
2 It acts in its future, in a similar way.

The action of such moments (parts of the cosmic process) in each other is conditioned by the fact that there is a maximum possible velocity of connection between different moments, *the velocity of light.* To illustrate this, we can draw a graph of space against time. For simplicity we consider only one dimension of space, which we denote by x. The maximum velocity (in two opposite directions) is represented by a pair of straight lines, as shown in the diagram, which pass through the point, P, representing the event under discussion.

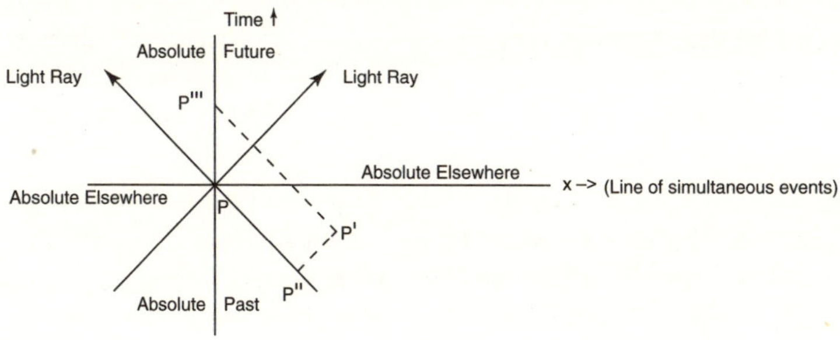

In relation to the point, P, the light rays passing through it separate space–time into three parts: (A) absolute past, (B) absolute future, (C) absolute elsewhere. The absolute past of P is constituted of all those events that can act in P. The absolute future of P is constituted of all those events in which P can act. The absolute elsewhere of P is constituted of all those events, such as P′, which have no *direct* contact with P. (Such a direct contact would imply a movement faster than light, which is impossible according to the theory of relativity.) Of course, *indirect* contact between P and points such as P′ in the absolute elsewhere of P is possible, because there can be other events, such as P′′, which are in the common past of P and P′, and which can be connected with both P and P′ by light rays.

The essential idea here is that any process whatsoever which can be localized at the point P can be acted upon only by what is in its absolute past. This is true whether the process in question is a part of inanimate matter or a sentient human being. Such a process does not fully determine its absolute elsewhere, but it does yield a *projection* of its past, which determines potentialities for its absolute elsewhere. Nevertheless, there is always room, logically speaking, for this absolute elsewhere to differ from any of the potentialities that are projected forward from the past of the given event, P. In other words, from the trace of the past of any given event, there cannot be a complete determination of the potentialities of the events that are simultaneous with it, much less of those that are in its future.

There is an instructive way of looking at this problem. Each event (including a moment of our own existence) is *presented* with that which acts in it. This is in reality what is *present* to it. The present of an event is really in its past. For example, what a person sees is the light now reaching him. With distant stars, this may have set out millions of years ago. There is no way at all to know what this star is doing "now". One may guess by projecting his own past forward to the star's position, but this will be only a guess, since it is always possible that something new will come into being between the events that we can now see on the star and those which come later. This discrepancy exists even for close objects, but it is usually unimportant for practical purposes, because these objects do not change appreciably while the light passes from them to our eyes. However, for very fast processes (which can be studied by instruments), this change can be of crucial importance.

We can see immediately that there is room here for objective and universal contingency. For once we admit that there is a *part* of the general cosmic process at some point, P, then it follows that the development of this part depends *objectively* on what is outside of it, since its future potentialities are contingent on what it is not.

Our own mental events are a special case of the kind of contingency described above. Thus, the inner structure of these events at a given moment determines completely neither the potentialities of what exists elsewhere at the same time, nor those of our own future. What I wish to stress here is that every event in nature has a similar inner structure (which is the trace, trace of the trace, etc. of its past), and that the further development of this structure is likewise not *completely* self-determined, but in some ways dependent on what is elsewhere to it. This is what I mean by objective contingency. This kind of contingency is essentially the same in physical events as it is in mental events.

Contingency must of course later turn into necessity, as the process at a given point, P, develops and further responds to the influences of the processes in its absolute elsewhere. This is the process of *necessitation* that we have been talking about. But there is also the process of *contingentation.* For certain developments, which would be necessary if only the projection of the absolute past of P were operative, cease to be necessary, because new factors have entered into the process from its absolute elsewhere. There is a continual process of necessitation and contingentation, exactly as you described in relation to viewing a picture. As a new experience comes in, that which was necessitated in a projection of past experiences ceases to be actually necessary. Room is created for something new, and as a result of the response to the very same experience, the new comes into being, its potentialities being necessitated by the experience in question. But as I have said before, this is also the way in which developments take place in nature generally.

It is clear that the above point of view emphasizes heavily the basically *inward* character of all processes, whether natural or human. As I have already indicated, in each "here and now", there is a moving trace of the rest of the world. This happens, of course, also in our own mental processes. Everything that we know is what enters into our inner process of consciousness at a given moment. The space that we perceive is a result of this process. Even what is outside is therefore really also inside. It is important to stress again that this is characteristic not only of human reality, but also of all nature.

The above represents a concept of existence as *action.* Each process exists in itself in the form of inner action, in a way that what is outside acts in it, and therefore also *is* in it. All relation is by way of such mutual action. Any part of the process, A, *is* for B, only in the way in which it *acts in* B. Thus, in effect, each part of the process "shows" itself by the way it acts in every part of the universe. It may act differently in different parts, and therefore show itself in different aspects. (This is really very similar to Whitehead's idea in *Process and Reality* that every event "prehends" events in its own past.) The basic concept is therefore *action.* Appearance is a form of action; viz., how something acts in the observer (e.g., in his eyes to form an image). Reality is in this way to be understood as *actuality*; i.e., a totality of action.

To sum up, nothing ever exists except some "here and now". Within each "here and now" are the moving traces which are the results of the action of all the "then and there's" which are in its past. Each "here and now" passes out of existence, to be replaced by others, but its trace is left in the inner movement and structure of subsequent events.

The commonly accepted idea of a single general picture of the whole cosmic process is wrong. Each moment gives, as it were, a broad "panoramic" view of its distant past. Processes near each other in space and time have a great deal of their distant past in common. From this, there arises the illusion of a single common "public" space–time. The only reality is, however, the set of "private" space–times belonging to the various events, and their relationship to each other. (Note that we do not fall into subjectivism, because the private space–time of the mind is just a special case of similar space–time is existing universally in nature.) We have here what may be called "concrete universality" of each event, since it is related to the whole universe, in a way that is peculiar to itself alone. For this reason, it is just as correct to say that the place and time of a given event (in relation to that of others) is *in it*, as it is to say that it is in its place and time.

It is important to notice that although each "here and now" has the *trace* of its past in it (and indeed *is* a trace of its past), it is not in general fully *conscious* of its past. For consciousness requires *memory*. Memory is more than a trace; it is the *re-creation* or *re-collection* of the past, as a kind of image of how it *was*. As you said in your letter, the trace is the past as it *was not*. Memory aims to produce again the past as it originally acted and to confront it to the action in the present. Memory involves a process of *reflection*, by which the inner trace is transformed into an *image*.

An interesting question is that of whether memory-like reflection does not occur to some extent in inanimate matter. The answer is that it does, at least in a rudimentary way. Consider for example the reflection of waves from a shore line, or even more obviously an echo of a previously produced sound from a rock face. In calculating machines, a similar process of reflection is frequently used to fulfil the function of storing up data for some time and then bringing it back at an appropriate stage into the process of calculation. In man's brain, such a reflection process is of course enormously highly developed. Yet it is clear that man's mental processes are not absolutely alien in this respect to those of nature. And as I indicated in the question that I asked you earlier, it would be surprising if they were. For then, how could man have evolved from non-living matter?

It is important to apply the ideas described above to ourselves as well as to nature. For, as is quite evident from the preceding discussion, we can learn about nature only through our relation with it; i.e., only through the way nature acts in us and we act in it. For example, a very young child may suppose that far away objects are small because they produce small images in his eyes. Later, he learns to correct this idea, by noting that apparent size depends in part on a certain relation of the object to himself; viz., the distance. But we have a natural tendency similar to that described above to weight heavily those factors that seem important to us, because of our desires, interests, fears, etc. We must understand ourselves and our relationship to the whole process in order to interpret correctly our experiences with nature.

One of the particular problems that must be understood especially well is the role of our own past experience, in determining our response to the world. It is clear that in this response, we are guided by a large number of ideas, conclusions, habits, etc., which have been accumulated. Accumulation is not a static affair, but a dynamic process. For that which has been accumulated has to be maintained. It tends to fall into a trace, so that a counter-process of reflection is needed to bring it back more or less to its original form. Thus, any set of well-defined ideas that we may have should be thought of as a system of inner movements. Let us call them "moving conclusions".

When we have a new experience, then in principle, we could respond in an integrated way to what may be called the "challenge" of this experience, by means of some action. Generally speaking, however, we tend to respond mainly with the aid of these "moving conclusions" consisting of old ideas, habits, prejudices, etc., in such a way as to produce another conclusion of this kind. Since these conclusions are only an infinitesimal fraction of our whole being, this response is not likely to be an integrated one. Rather one part of it will tend to come into conflict with other parts thus setting up a process of disintegration, which in general becomes apparent only later, when it begins to act in nature, in other people, and in ourselves.

Whenever there is such a lack of inner integration of response, true creativity is impossible. Such creativity can come into existence only when the set of "moving conclusions" described above is de-necessitated, or in other words, when the process of continuation of these aspects of the past by means of a corresponding movement of reflection comes to an end. This cannot be brought about by trying to oppose, destroy, or annihilate these aspects of the past. For as you point out in your articles in *The Structurist*, this only perpetuates their dynamic power in another form. Rather, it must be done by trying to *understand* the past. If this understanding is deep enough, we are liberated from the domination of the past; we stop repeating it mechanically or with mechanical modifications, and we are free to respond creatively to the challenge of the new moment.

In this connection, I regard work such as yours, which analyses developments in the field of art, as being of crucial significance. By seeing into the process of art as it has actually been, we come to understand it. In this way, we cease to be under the necessity of repeating previous trends within the older general framework, or else of simply trying to destroy the old framework altogether, thus still being conditioned by the past.

Nevertheless, just because I think that your work is so important, I feel the need to express certain misgivings, which arise in me when I think about some of the things that have been written about Structurist art. In other words, I have the feeling that along with a great deal of truth, the theory of Structurism may have some inconsistent elements. These could impede you in carrying out what I feel to be your real goal, which is to help us to an understanding of the role of creativity, first through art, and perhaps ultimately in a more general way.

As far as your analysis of the limitations in past trends in art is concerned, I am in general agreement. When you suggest Structurism as a *possible* way out of the impasse, I can only say "More power to you". The difficulty is that your theory has in it certain potentially dogmatic elements. For you feel that in addition to studying the facts of art as it did evolve, you can pick out the real main trend, which Cézanne and Mondrian carried along without developing it to the full extent of its possibilities.

Now you may well be right in your assessment of this trend, but you may also be wrong, or only partly right. The precise truth about this question is however, much less important than is the fact that while you project this trend towards the future, it *is* actually only the past, as seen from your own special vantage point. Other people also look at trends in the past and pick out a certain one that appeals to them, which they too think is the proper thing to be developed.

The question that I want to ask you is this: "Is not the pre-judging of the issue of what is the main trend in art incompatible with full creativity, which must be free from moment to moment?" I realize of course that you are allowing yourself a great deal of freedom, but only within the framework of the Structurist theory. Suppose that in a truly creative moment, you were led to something that conflicted with the whole set of basic tenets of Structurism? Are you allowing yourself to be free to do this, if the challenge of the future moment should lead to such a response? Or are you perhaps unconsciously restricting your responses so as to avoid such a possibility?

I feel that I must clarify a few points here. As I said before, I agree with you that true creativity is not a simple destruction of the past. Nor is it a simple repetition or continuation with modification of the past. It must be free to reject any feature of the past, to continue it, to modify it, or to initiate some new line of development which is not limited by the past. This freedom applies not only to what other artists have done, not only to the work of that particular artist himself, but also to all the theories that this artist may have made. For these theories are themselves based on the past, and *are the past* of this particular artist, even if they are his assessments as to what he imagines that the future will be. Such assessments are generally incomplete in two respects. First, because of his own particular character and background, he will be attracted to certain aspects of the whole process of development, repelled by others. These attractions and repulsions are arbitrary. Therefore, he is in danger of picking out the tendencies that attract him, and calling them the true continuation of the process. Secondly, there is (as happens in nature too) always something new coming in from outside, the unexpected. A dogmatic adherence to theory may prevent him from being sensitive to this.

As I see it, the purpose of theoretical analysis is to *free us from being controlled by the past that is in us* (i.e., the "moving conclusions" that I mentioned previously). Such freedom cannot be obtained by ignoring the past, because the past is always operating in the present, whether we are aware of this operation or not. The artist (and of course the scientist too) must therefore try to understand this past, not only in the way that other people are influenced by it, but also in the way that he *himself* is thus influenced. (For example, he must see that his own theories may be conditioning him in the same way that other people's theories are conditioning them.) Until one understands the true role of the past, as it works in one's own mind, one tends to over-emphasize its importance. As a result, one is incapable of an integrated response to what is new in the present.

In this connection, some of the things that have been written about Structurist art resemble what the Marxists have said. For they felt that by studying the evolutionary process of the past, they could pick out the main direction in which history was moving. They became so attached to their theories that they were unable to review their own role objectively, or to admit new and unexpected developments not fitting into these theories. It is interesting that they too presented the argument that a partisan viewpoint is inevitable. One must be conscious that one is a partisan and take it into account, they said. But they were never really able to do so. Perhaps this was not entirely an accident, but partially a consequence of a theory that extolled partisanship. In other words, they not only admitted that they were influenced by what attracted and repelled them, but said that this was only right and proper, and indeed a wonderful thing. For without it, they would be unable to be active enough and ruthless enough to carry out the revolution which their theory demanded. To recognize partisanship as something that *is there inevitably* is one thing; to say that it is desirable is something else. For example, one may recognize that disease is probably inevitable, but if one extols it by regarding it as a mechanism for killing off weaklings and thus improving the race, one is introducing a basically false idea.

In this regard, I feel that Bornstein's article in the *Structurist* that you sent me sets forth with particular clarity the aspect of the theory that I am referring to above. He says several things that could equally well have been said by Marxists. For example, "Without the idea of progress – meaning a conscious moving on towards something worthy of faith and endeavour – there is no history." Elsewhere he states, "Faith is commitment." He then refers to the fact that your work has revealed an order or process through which man's art has evolved, an order which presents a positive direction of advance along a new course. The Marxists have said something very similar about their own point of view. Suppose now that a person is confronted with two conflicting faiths. Each of these faiths is equally honest in recognizing its own partisan character, and in trying to take its own subjective limitations into account. Since it finally comes down to a question of faith and to commitment to what is in the nature of the case unknown, i.e., to what will develop in the future, each person can finally say only that the faith of A is what appeals to him while that of B repels him or vice versa. Does everything really finally boil down to such accidental and arbitrary personal preferences? I do not believe so, nor it seems, do you. For if it does, then your whole position, which is against Tachism, Action Painting, Expressionism, etc., becomes untenable. Indeed, all these finally put the essence of art in the accidental personal idiosyncrasies of each individual.

It seems to me that what you are most basically concerned with is the role of creativity in art. Analyses such as yours may well suggest new lines that are worthy of being explored. But unless this exploration is carried out in an undogmatic way, without "faith" or "commitment", then it seems to me that you are in danger of interfering with the freedom of the creative process. For you are not allowing yourself to carry out such a process, if it has aspects that are not compatible with the articles and commitments of your faith.

It seems to me that true creativity requires that theories be regarded as dispensable hypotheses, generally useful up to a certain point, but not to be taken so seriously that they should channel the creative process. As I indicated previously, this does not mean that the past is to be ignored, or that one is to fall back into mere subjectivism. Instead, it means that the total situation in each new moment can lead to a challenge requiring an integrated response. This total situation is produced by nature (and of course other human beings) as they impinge on a given person, and also by the further inner development of that person himself, by which he understands his past and frees himself from its domination. It should be clear that I do not regard Expressionism, Tachism, etc., as true creativity, since they stress too much the immediate and accidental, because not fully understood, subjective drives of the individual artist.

I think that the above few pages express the essence of my present views about Structurism. It is important, I think, that we discuss these, as they ought to be cleared up before I discuss your position in more detail. I am therefore awaiting your comments on these points (as well as on the rest of the letter of course) with great interest. Meanwhile, before receiving your answer, I shall probably send you another letter discussing some of the additional interesting and important points that you raised in your last letter; viz., the role of contradiction, order and disorder, process as a unity of opposites, the role of emotion in thinking, and finally, the limitations on expressing reality in terms of words and their associated thoughts.

Very sincerely yours,
David Bohm

P.S. Enclosed are some clippings about the exhibition of Structurist art that I attended in London.

Red Wing
Route 2
Minnesota, USA
March 2, 1961

Dear Mr Bohm:

Thank you for your very good letter, I appreciated hearing what you had to say.

You ask me: If error and contingency are peculiar to human life? But if this life evolved out of general nature on which it remains dependent, how then did humans attain the above qualities if not from the nature out of which they evolved?

The answer is in those qualities which uniquely distinguish human life from *all* other aspects of nature, both physically and mentally. Physically, man has attained the upright position which has the important consequence of freeing the two front limbs from the earth as well as the digits on those limbs, into incomparable areas of achievement. Mentally, there is the immense expanse of human consciousness, both qualitatively and quantitatively, with the possibility of indefinite future expansion, for only a small part of the brain has been used so far.

How then to *relate* these *distinctions* to their origin in extra-man nature from which man evolved? We speak of distinctions within the area of human life, such as primitive in contrast to so-called civilized man. Such distinction-relations are also extended to inanimate nature, as in your example regarding memory. In man memory is a part of consciousness, but when "memory" is traced to its lowest order in inanimate nature, there is no consciousness. If a stress is exerted on a piece of plastic and afterwards it returns to its former state, it does so because the applied stress did not destroy the persistence of the original order. What a world of difference between man and the inanimate. If I see a shell, I react to its image. The shell, however, neither sees me nor can it react to me. The shell remains in my memory, I never enter the shell's "memory." If this sharp distinction is not true for non-human organic life, the presence of these qualities are strictly limited beyond which development ceases.

The ability to abstract is the most potent center around which the conscious life of man is focused. Other organic life evidences this ability. But in every instance there is a limit. Only man has unabatedly continued to extensionalize his abstracting ability, that is, develop his consciousness of nature and self. For this reason he alone produces science, art, in short, culture.

You can see the line of my thought without further elaboration, that is, to distinguish human life while, at the same time, relating the distinctions to nature. There remains to consider why humans, being supreme in the realm of consciousness, that they alone must resolve problems of contingency?

Human life evolves in the interior of a female, from the most elementary cell structures through all kinds of suggestive primitive forms of sea and land organic life. At the last moment this life sheds its completely hairy body, and is born. Biologically and historically, there is an evolution of human life out of the original state of nature.

We think of the evolution of the universe. We can also think of the evolution of nature centering around a conscious form of life. Nature, of course, would continue if man were to disappear tomorrow, now a possibility. Nevertheless, indefinitely expanding conscious life introduces a unique factor. It must seek comprehension not only of the endless ramifications of nature structure, but also to comprehend the ever-expanding ramifications of its own desires within the structural potentialities permitted by nature.

Extra-human nature continues without consciousness; human nature must ceaselessly improve its consciousness of abstracting from unconscious nature processes for the survival purposes of a consciously directed existence. Nature is like the plastic structure mentioned earlier, while man has the double task of comprehending unconscious and conscious aspects of nature. Now consciousness implies choice. Therefore, what lacks consciousness cannot evoke the act of contingentation. What possesses consciousness, especially human life, is literally surrounded by choice and, therefore, contingency exists on all sides.

If the above holds, attributing contingentation to non-human nature would appear as anthropomorphic projection, as also in the instance of inanimate "memory." The latter is a purely physical manifestation of structural action. If we take the physical uniqueness of the human, its unique expanding consciousness, in short, its unique capacity to physically–mentally abstract, and follow its "traces," the quality of contingency disappears in organic life at a point, while the last remnants of consciousness and abstracting end with the ameba. The further back we go, the more necessitation assumes the dominant role until it overcomes contingency actions altogether. For example, inanimate "memory" is seen to be purely necessitative. The moment you leave man you have left all forms of life that are necessitated to comprehend nature in terms of itself. Beyond organic life there disappears even the necessity for anything to know anything, in contrast to man who strives to know everything. This situation makes contingency unique to human life, forming the bridge or the abyss, to make possible the survival or the extinction of human life in nature.

Man is not like a rock, tree, bird, animal, etc. Such things fit into patterns before which they have no choice but to submit. None of these things can invoke change, all are prisoners of the general process of nature. Only man has become a free citizen of nature and can become more free. (At present man is largely a prisoner of human nature.) For this reason man alone is able to introduce change in nature, including his own nature, either by carelessly tampering or creatively seeking harmonious relations with nature. In the latter he puts into action aspects of nature structure, which are latent in but so far untouched by nature itself.

Human life thus introduces the factor of contingency into nature on two related levels. By seeking comprehension of a nature that gets increasingly complicated, and seeking to work out his own volitional existence within nature, a goal that has also become immensely complicated.

I understand you to say, that nature consists of a number of processes, each distinct from but related to the other, with each process possessing its own essential character which it retains regardless of what happens in the course of time. If this were not so, we could not recognize what you call traces, and traces within traces, etc. Nor would nature consistently produce specific structures, on all levels from particles to the macro-level. Each "event," as you say, has a unique character. Strict individuality reigns everywhere in nature, is what I take it you are saying. There are only "private space–times," you write, each of which is but a special case of the general space–time of nature. Therefore, each event is a manifestation of what you call "concrete universality," each individually related to the general process of nature. Does not this imply that each atom, particle, possesses individuality, contrary to certain physicists, the question I wrote about in *TS*? Can we not say that man confronts an ambiguous experience of nature, his task being to ascertain the unique individualities of events which are obscured by the ambiguity?

What you say about the role of past experience is well put, indisputable, and of first importance. We easily become conditioned to our past in place of the conditionality that is necessary. We are in complete agreement that canalized orientation to our past frustrates creative action. You then come to express your misgivings about my art attitude, expressions of which have appeared in other of your letters. You wonder whether I have sewed myself in my past, plunging ahead with dogmatism. I would not claim that I am innocent of dogmatism, error, stupidity, etc. The question is, how serious they are. In this respect you would have been more helpful to me if you had applied your theories more directly to my actions, art or art theory. I invite you to do so.

You ask, whether "prejudging" the art future is compatible with the effort to be creative? Is it possible to avoid doing so? We are not so constituted that we can hold ourselves in abeyance until each moment of the future appears. We have to prepare for it. We can no more ignore decisions about the future than of the past. It is rather a question not of whether, but of how. Our entire correspondence, as I understand it, deals with the solution. It is not that we do so and so with the past or future in order to preserve creative action, rather we begin with creative action to determine how we will deal with the past or future. Even if the general aspect of my prediction of the art future holds, I know it to be a foregone conclusion that the actuality will be greatly different. If this were not so, the future would contain no surprise, no discoveries to be made, and no revitalization of the general art effort. If the artist strives to avoid such predictability, he will be unprepared to use the future events. Thus the spectacle of artists applauding the virtues of figuration today, of non-figuration tomorrow, and then once more around. Since the event of artists claiming the prerogative of creator, whether falsely as Picasso or not, he has automatically taken on a number of new responsibilities. Nature no longer protects the artist.

There is a more fundamental reason why we must prepare for the future. Past man could live in a state of evolution in which natural events more or less maintained human survival. Art and science had their models of study supplied them by nature's creations, and with them secured an evolution. The past was largely natural or unaided evolution, it was not too necessary for man to be conscious of his historical process. Around the close of the nineteenth century all this was altered radically. It was evident only much later when science was found to have gone so far in its manipulation of nature creation, as to make human annihilation possible. Artists confusedly sought to be creators, and not just follow the creations of nature.

Man's evolution had reached a point where he sought a dominant place under the sun. What is not widely recognized is that this point in his evolution and ambition automatically brought with it entirely new responsibilities for future evolution, if he was not to perish by his own hand. That is to say, man no longer possesses the natural protection for his evolution, he is now involved in the direct manipulations of basic nature, he must decide where he wants to go. If man's present manipulations of nature are left to the mercy of natural events, destruction will result.

The past is largely unconscious evolution, first physically and for a while mentally; now the problem is to achieve conscious evolution. Even if there were not the threat of physical destruction, there would still be the threat of mental destruction. How then can we possibly avoid making decisions about the future? The danger is in absolute decisions as the dictators are in the habit of doing.

In the face of the dictators it is disastrous to avoid at least tentative decisions about the future. It is now necessary to be as active in our preparation for the future, as we once were only with the past. The alarming thing to me is that only the dictators are exploiting the new possibilities of the future. Their success, however, is due to their use of the most primitive means, brutal force, and the most developed corruption of civilized man, brute falsehood.

Am I in a position, you ask, free to recognize anything that might challenge the basic tenets of Structurism? Am I unconsciously shielding myself from seeing such possibilities? This, it seems to me, you will have to decide yourself. Otherwise, I can only say what I do. Two things. The first has to do with myself. All my theories are put aside, I am alone with my art, my experience of it. I cannot give these experiences, others must get them. If I could give them the art would be expendable, there would only be a form of art literature.

The other thing has to do with other artists. You cannot know life and ignore death, health and ignore disease, truth and ignore falsehood. I try to comprehend all the major art attitudes prevailing, to literally project myself into the artist who made the art. I thus concern myself with what is opposed to my view. If the artist fails to do this, it will avail him nothing to be well versed on the dangers of self.

Regarding the similarities between Marxist and Structurist history theory. The Structurist theory, as I offer it, is taken to be the most fruitful course open to the future of art. If the Communists win tomorrow, this art would disappear. This art is not inevitable, nor would it succeed in its own goals if forced on any artist. To the contrary, it is both the old and the new status quo who would suppress it. Communism is the old cloaked over with the new. To make a comparable case with Structurism, it would have to advocate some such art as Picasso, whose attraction to Communism is one with his art. Structurism does not bring in the old disguised with the new. I cannot answer for Bornstein's quotes, not being responsible for them. Nevertheless, I concur completely with your criticism of them. Structurist art was evolved precisely because of an unwillingness to accept or reject art on the basis of one's urgency for commitment or one's capacity for faith, the prevailing practice since the demise of mimetic art. For this very reason I reject Mondrian's Neoplasticism which, as the artist himself says, replaces religion.

You close your remarks on this subject with warnings of the dangers of theory. In *The New Cézanne* I had occasion to discuss this subject several times. I ask you to read in one of these places – the last two paragraphs on page 14.

Music: closer to mathematics, hence the physicist's preference for it over art. What do I think? No great composer has used mathematics. Why is this so if it "has a more clearly defined relation to mathematics?" I suppose you know Brikhoff's work on math and music.

Since the disruptions of both art and music in the nineteenth century, many individuals have attempted to compose with math. Nothing remotely convincing has yet resulted. A clue to math's appeal I think appears in the more recent tendency to raise the banner of indeterminism in the arts, with or without the alleged appeal to mathematics.

Thank you very much for sending me the reviews of the English Constructionists. I would not have seen them if you hadn't taken the trouble.

I welcome your criticisms of my Structurist art or theory. Whether I agree or not, I will profit from what you have to say.

With my best regards,
Charles Biederman

3, Berkshire Road
Bristol 7
Glos., England
February 24, 1961

Dear Mr Biederman:

I am now continuing the letter that I have already sent to you, in order to try to get some of the many questions that you raised in your last letter into better order.

Perhaps we ought to begin on the problem of contradiction. In your letter, you raised questions which go very deep indeed. Of course, you are right when you say that in terms of the framework of a given set of concepts, we can describe reality only within certain corresponding limits. Or as you said: "The verbal statements of realism only indicated the conscious condition of human vision (reflection) of nature at the time." All of my discussions are, however, aimed at the attempt to describe reality with the aid of definite *terms*, which may include words, symbols, pictures, etc., and the thoughts that they stand for. I wish to assert:

1 As long as we use such terms, there will have to be conflict, contrariety, and contradiction *in our description of reality and in our thinking about it.* (With this you apparently agree.)
2 The use of such terms is not *just* our own invention. Rather, in some sense, something analogous exists in nature. (This, you do not seem to agree with; so that I will have to try to justify it.)
3 It is because of the analogy between terms in our thought process and something in nature that is at least similar that our procedures of thought are able to have enough correspondence with nature to be useful.
4 The problem of going beyond the use of terms is a very fundamental one. Indeed, as I shall try to explain later, if we could do this, we could live in a continual creative relation with nature and with other human beings.

Let me begin by defining the word "term". This has the same root, of course, as does "terminate" and "determine". What we mean by a "term" is a *conclusion* of a certain thought process. Of course, the thought process is, like all process, without limit or measure, at least in its general features. However, when we reach a suitable conclusion, we *terminate* it. This does not mean that we bring it to an end and then forget about it altogether. Rather, it means that we stop its natural development of falling into a trace of a trace, etc., and try instead to make it reflect back, so as to return continually more or less to its original form. In doing this, we set up a "moving conclusion" of the kind that I referred to in my previous letter. Such a "moving conclusion" is then added to the whole set already acting in our memories, habits, presuppositions, etc. It is at this stage a part of the bag of tricks with which we tend to confront the world. In particular we use such "terms" in our thinking process. It is not an accident that our language refers to "thinking in terms of such and such". Indeed, without any terms at all, how would we ever think?

Now, as we both agree, process cannot be fixed in any definite form. I like to give the analogy of the lights of Paris reflecting from the Seine River. When one stands on a bridge, one sees flashes here, there, and everywhere in the water. These flashes are like our "terms" of thought and discourse. They evidently introduce conflicts and contradictions that are not in the city that they are reflecting. Any attempt to think in definite "terms" is evidently capable of introducing conflict and contradiction in a similar way, not only because nothing can be fixed in a moment of time, but also because the structure of each moment is (as we both agree) the result of the whole process by which it becomes what it is. Since the effort to describe the process with the aid of terms leads to contradiction and conflict, so also must the effort to describe the structure that is present in each moment. In other words, as long as we try to use "terms" to think about the world, to communicate with each other, and to record the results and conclusions of our thoughts, we must expect contradiction and conflict to appear quite necessarily, since we are not thinking about the world *as it is*, but rather, *as it is not.* The really interesting point, however, is that, *to some extent at least*, process can be reflected adequately with the aid of terms, *provided that we admit contradiction and conflict.*

All of this is, in a certain way, just common sense. Thus, it is evident that the act of speaking words and thinking thoughts constitutes a process, which is taking place in the real world. Nevertheless, while words and thoughts are real processes, the actual process of the world is not constituted solely of words and thoughts. It is therefore hardly surprising that the effort to express reality by words and thoughts must involve contradiction. The most important question is, however: "In what ways and to what degree do such words and thoughts constitute an adequate reflection of the overall real process?"

As a first step towards understanding this problem, let us ask what is meant by the term, "contradiction". As its root implies, this word means to "speak contraries" or speech that goes against itself. Is reality contradictory? My own view is that reality is neither contradictory nor non-contradictory, but rather, if I may coin another word, that it is simply "non-dictory". Or as I stated earlier, it is not to be understood solely with words, or with terms of any kind whatsoever. However, since we have no choice but to use words and terms in our thought processes, let us try to see how far we can get toward the *implicit* character of the reality that is at the basis of the terms that we use, the reality that *we really mean* when we are talking about it or thinking about it.

The first question to be considered is whether there is anything in the actual process of the world which is analogous to the terms that appear in our thought and speech. If a given process is analogous to another, it means that one of the processes contains some parts, A, B, C, etc., which stand in the same relation to each other as do certain corresponding parts, A′, B′, C′, etc., of the other process, of which it is an analogy. Of course, no analogy is perfect in *every* respect. Nevertheless, some analogies have such a wide range of correspondence that they afford considerable insight into the problem involved.

Now, the first application of the *process of terming* is in our experience of objects. When we recognize an object, we give it a *name*. Eventually we recognize that it is one of a class or set, and give it a more general name. Thus we proceed by progressive abstraction, to include objects in broader and broader sets, each of which has a name or a term.

What justification is there for giving names to objects and classes of objects? First of all, we do not assign such a name to a process seen to take place in the world unless it has certain characteristics; viz., regularity, repetition, stability, and similarity to other processes that are by its side or that are stored in our memories or in other records as traces. It evidently must have a recognizable form and a fairly well-defined place (in relation to other objects). It must also have other persistent qualities and properties (colour, hardness, inertia, etc.). Thus, in this context, a "term" implies a related set of persistent and stable attributes of the object that is "termed". Therefore, *the stability of the term in our thought process is not something that we have completely arbitrarily introduced out of nothing*. Rather, this stability provides an analogy to a corresponding stability of the attributes of the object. In other words, we have introduced into our thought process a feature that is analogous to one that is already present in the process of the object.

It is important to point out here that as a rule we do not *notice* the remarkable similarity between the way the *thought process* treats this problem and the way it happens in the *process of the object*. This is because we do not usually think of thought as a process. Rather, we tend to identify thought with the *result* of this process; e.g., as written in books. But if all men were destroyed and their books survived, there would be no thought. Thought is an *activity*. The books are both a result of this activity and a structure helping to determine potentialities for further activities of a similar kind. If we look at the *whole process* of thought and not just at its end result, we see that there is a continual movement and counter-movement, even in the most elementary mode of thinking. When we assign a "term", we initiate a process of reflection and repetition, so as to bring the movement back to its original form again and again. But this is just what is going on in the object too. For example, at the physical level, its molecules are moving back and forth, its waves pulsating to and fro, etc. Thus, the relative permanence and stability of the object in the overall process of reality is of the same general nature as is the relative permanence and stability of the corresponding "term" in the overall process of thought. (Note the analogy: "Object is to overall cosmic process as term is to overall thought process.")

It is evident that an object is named or termed with the aid of its characteristic qualities, properties, and other attributes. (We include place, time, and form, among these attributes.) A quality or a property is evidently not a concrete object as such, but rather, an activity or function, which the object is able to carry out under appropriate conditions. We generally begin with functions that the object stimulates in ourselves, through our sense organs. Thus, an object exposed to light will stimulate in us (through our eyes) certain functions by which we recognize place, form, colour, etc. If we feel it with our hands, the functions will correspond to its texture, hardness, etc. A particular name for an object is always associated in our minds with a certain combination of such functions that generally appear together, whenever we sense the object in question. Vice versa, this combination (or some significant part of it) can elicit the process of naming it.

But once we get this far, we see that very much more is implied than the mere factual description that has been given above. For there is a whole range of possible objects that can stimulate the same function within our minds (e.g., the function of seeing the colour "red"). The function therefore *selects* objects in accordance with those that stimulate it and those that do not. This selection first of all *separates* objects according to their different functions, but secondarily, it *unites* those objects which have a given function in common.

The above separation and union of objects into sets and classes is the starting point for our ideas concerning the particular and the general. Each particular object is defined as that one which stimulates in us every one of a certain set of characteristic functions. (This definition is evidently *implicit*.) As we abstract from some of these functions, we are left with a smaller set, which can be fulfilled by more objects. Thus, by abstraction we proceed from the particular to the general. By associating names to sets of function, we have built a certain *structure* into our thought process; viz., the structure which is at the basis of the use of sets of particular objects and sets of sets etc., with ever greater generality.

Let us now see how these sets enter into our thought process. Every time we see an object, we must see a certain set of qualities (functions within us), which finally elicit the naming process. The name then elicits more of the set of associated functions. But each such function is associated with many objects. Thus, if the process is allowed to go forward, it will lead to another object, another name, still another function of the object, etc. In this way, we get the string of associations, the "stream of consciousness" that is characteristic of uncontrolled thinking. But then, if we wish, we can *pay attention* to what we are doing. We no longer allow thought to develop idly in a random stream as described above. Rather, by a conscious effort, we can *reflect* the thought of the function back to the same object, then allow it to fall back to another function of that object, then back to the object, etc. Such attention is not just a passive exclusion of other associated objects. Rather, it is an *active* (and frequently arduous) process by which the thought is made to reflect back on itself. This is what I meant previously when I said that even to think of the most elementary "term" is an active process of reflection (from name to function and back), which resembles in its general features certain of the processes by which the objects themselves exist.

Let us now go more systematically into the organization of ideas by association. Now, just as an object can elicit various functions in our minds (e.g., attributes, qualities, approval and disapproval, pleasure and pain, etc.) *the functions themselves* can be treated as objects. For example, qualities, such as redness, greenness, blueness, etc., can be given names, and one can attribute qualities to these functions (e.g., warmth, depth, etc.). These second order qualities are also functions in the mind. In this way, the mind begins to reflect on itself. This will produce a further kind of ordering of thought. Thus, just as each of the first order functions united certain sets of objects, and separated others, so each of the second order functions will unite and separate the first order functions. But now notice that an object is represented by the union of all the first order functions which it elicits. As a result, *the second order functions will be able to represent objects*, since each one can unite the first order functions, qualities, with a group, corresponding to a certain object.

This means that the *mental object* is a second order function. The mental object is what arises when we remember something by association. It is constituted by tying together a whole set of quality functions, by means of a single second order function. (It is possible that memories can be recorded more directly, as a kind of "photograph" or "tape" of the past, but memories in such a form will be difficult to recall, as they will not be tied in with the web of associations, which is at the basis of our "everyday" thinking.)

The essential point in the above is that an object is treated in the thought process as a function of a function (i.e., a second order function). Of course, one can go on to third order functions, which were elicited by mental objects (e.g., memories go on to produce changing responses). The main point I wish to make here is, however, that everyday thinking is mainly the response of memories. Thus, when we see something, certain qualities are elicited functions, elicited in us, and these set up some object function. We then compare our immediate perception with the mental object functions. Either we see that it seems to be the same as the mental object, or that it differs. If it differs, we say it differs in certain *respects*, in certain *ways*, or in certain directions. (For example, it has *more* of one quality, *less* of another, etc.) In order for such a comparison to be carried out, we must have a further mental function, i.e., *the function of comparison.* This function makes it possible to assert *difference and identity.* Without such a function, our memories would evidently be quite useless and indeed practically meaningless.

Now it seems to me that the function of recognizing difference is more elementary than that of recognizing sameness or identity. Thus, both man and animals can frequently tell that something has changed without knowing precisely what is there in the first place. There seems to be an elementary sense of change or difference. "Sameness" is then taken to mean "non-difference". That is, let us suppose there is a certain process of registering difference or change (e.g., this moment differs from a previous one). The non-functioning of this process would be called the registering of identity.

No other sensible understanding of identity seems to be possible. If you think otherwise, why not try to give another interpretation of sameness?

The above interpretation of identity helps throw some light on our use of language. Thus, we frequently assert that two different objects or qualities are the same (for example, that two apples have the same quality of redness). If the two were precisely the same, then it would be impossible to assert their difference. What we mean is that if A and B are identical, *they differ by nothing.* But the word "nothing" has here a relative meaning. It means only that the differentiating function in our minds does not work.

The notion of a relative nothing, or "nullity" is quite a common one. Thus, we may say "This is nothing to me", meaning by this, not that it does not exist, but rather, that it has no function within the sphere of my interest. In mathematics, the notion of a "relative zero" or nullity is well known. A certain symbol may function as nullity in a given field, but not in a broader field. Thus, every equation should be written not as A=B, but rather, as A–B=10?, where 10? represents the relative nullity of a given field. Indeed, it is fruitless to use two different symbols A and B, if both sides of the equation represent absolutely the same thing. But if A and B are different, then the equation makes a real contribution to our knowledge, for it asserts that their difference, whatever it may be, has no effect in a certain field.

In physics, there is a similar relative nullity. Thus, in the field of measurements of a certain sensitivity, effects below the "threshold" of sensitivity of the apparatus register as "nothing". And in the whole of nature and human life, the world is full of such "thresholds", such that what is below or beyond them simply does not function in the field under discussion, so that it is represented in that field as "nothing".

In the field of the thought and experience of a given person at a particular moment, "nothing" refers to all that has no functions in this field. Of course, what was "nothing" at an earlier moment may function later, either because of the changing relation of the object to the person, or because the person has meanwhile become more sensitive and perceptive. Thus, the word "nothing" is always changing its meaning. At each moment, it means the infinite and unfathomable sea of reality that lies outside the field of my thought, my consciousness, my experience, my memory. But from moment to moment, different aspects of this sea will enter my field (as will be true of course for every human being). In other words, the total field is the field of all my thought, all my experience, etc., *plus* the nothing of this field. I must recognize that my thought, experience, etc., while a part of this reality, is not the whole, and that the object of my thought can be represented in the field of this thought only by the thought of nothingness. That is to say, the object is not a thought at all, but it is what is projected into the field of thought.

Now we can come to the problem of contradiction. Every word, every thought, aims at some generality in its meaning. For example, if I see an object today, I notice its roundness, redness, size, etc., etc. To "place" the object, I compare its qualities to other qualities in my memory. Thus, I say that its "redness" is the same as my memory of redness. But evidently this is not really true. Every experience of redness is different. Today's redness cannot be yesterday's. Therefore, if I assert that two objects are identical in being red, I am involved in contradiction, since there cannot be the *same* redness in different objects. At most, I can assert that their redness has a null difference in my own field of thought and memory. That is to say, when the immediate perception of the object is compared with memory, the "difference function" in my mind does not operate with regard to the quality of colour.

A similar problem exists in physics. Thus, if I assert that this stick has the same length today as it had yesterday, I cannot go back to yesterday, to see how long it was then. But I can only compare its length to the length of some other object present today, which is a *trace* or *memory* of the stick of yesterday. And in this comparison, having the same length means only that in a certain field of measurements, no difference is recorded.

Now the whole utility in language depends on the fact that words have a *general* meaning. In other words, we can use the terms redness, roundness, weight, size, etc., in an unlimited set of contexts, and mean the "same" thing by such terms. If we had to use a different word for every different experience of "red", then really we should not bother to use words at all. As we have seen, however, it is wrong to fix the meaning of words, for in reality, they never have the same meaning. Therefore, every time we make a definite assertion, there is a contradiction. We assert that the word has a general meaning, and yet we must also assert that it does not; i.e., that it is different in each particular case.

We could get out of the contradiction by asserting that identity means "differing by the nothing of a certain field" (e.g., the field of all my experiences up to a given moment). But then we have only pushed the problem into the word "nothing". For it no longer has any well-defined meaning. Because the word "nothing" is now so flexible, we have no assurance that the redness of one moment is the same as that of another at all.

What I propose is that the way out of the trouble is to recognize that contradiction is inevitable in the movement of thought and language. Each thought really has contrary meanings (hidden in the infinite and unfathomable sea of "nothingness" of thought). And it is essential to allow such contrary meanings. At one moment, one meaning is emphasized as dominant; at another moment, a different meaning. But the communication of the overall meaning (i.e., the really that we really mean) is through the ebb and flow from meaning to counter-meaning. I may take a comparison here to counterpoint in music, where the theme is carried in reflected and opposing ways in different parts (treble and bass). So perhaps it should be called "counter-diction" and not "contradiction". Then I would like to propose a little dictum: viz., "no diction without counter-diction".

I would reserve "contradiction" for another process; viz., an *unrecognized* counter-diction, whose outcome is *conflict*. Thus, if one does not recognize the ways of thought, one will not know that the secondary and suppressed meaning will continue to operate in the secondary process, to have an important effect, which can eventually come into conflict with that of the primary meaning. For example, even in the field of measurements of a given accuracy in physics, the effects of errors below the "threshold" of sensitivity can accumulate, to produce visible effect above this threshold. (For example, if the minimum threshold in a single measurement is 1 unit, then after many measurements, a series of errors of 1/10 unit, each of which registers as "something", so that a series of "nothings" become a "something".) Similarly, in life, the contradictions, each of which is outside the field of what we have experienced or thought of, can accumulate, to produce a result in the field of our experience, a result that generally conflicts with the effects that we aim for with the part of the process that is known to us. We must therefore be continually alert to the existence of counter-diction in our thought process, which can give rise to contra-diction or conflict, when the conclusions of our thought lead to action.

You may ask, "Can the mind not be free from counter-diction?" Evidently, this is not possible as long as it functions in the field of thoughts, which are associated with words or "terms" that attempt to fix the thought process in certain ways. Can the mind be outside this field? I think that the answer is "Probably – yes". Perhaps in creativity it is this way – it may have a total understanding of its whole field, including even itself – without a division into subject and object, "I and not-I", past and future, etc.

All of these divisions are in the field of counter-diction, since they are established by functions, such as those of difference and non-difference (identity). Perhaps when the mind is outside the field of counter-diction, it can see the real meaning of these divisions, and, as it were, "project" certain conclusions into the field of the divided mind. This latter field is necessary to translate the creative process into action, but the creative process may perhaps transcend the field of time, in the sense that it understands the order of things in a deeper way, which shows why actions taken in the field of time are subject to the law of counter-diction. If it were possible to see the world and oneself as a totality, undivided (which is perhaps what happens in true creativity) then the need for counter-diction in the field of everyday life and the process of transcending it would perhaps be immediately perceived.

I would appreciate your ideas on this subject. Incidentally, about the relation of mathematics to music, which we discussed earlier, it has long been recognized that these are closely connected. Thus, in ancient Greece, the Pythagoreans got some of the inspiration for applying the concept of number to the whole of reality by noticing the simple ratios of the lengths of strings that are in harmony. Much of the structure of music is similar to certain features of mathematics. The scale is based on *number*. Then there is rhythm, which is also a highly numerical concept. There is the problem of various kinds of time in music. Then the repetition of a certain pattern of notes at another pitch, in another key, etc., is strongly reminiscent of group theory. Also, the theory of reflected patterns, opposing patterns, etc.

P.S. I forgot to mention that the practical utility of the concept of the *general* meaning of a word depends on the fact that in nature, there are *fields* in which certain kinds of difference do not function (i.e., make no difference). To the extent that such fields exist, we can utilize the mode of thinking, in which different individuals are treated by the "same" set of qualities and attributes. In other words, if A and B differ by nullity in their effects in a certain field, they can be called "identical" in that field. The contradiction arises because when the field is studied more closely, or if one allows a process to develop in time, "null" differences can lead to something that is "non-null" in the field, while non-null things can lead to nullity. Thus, "nothing" can become "something" while "something" can become "nothing" – which is clearly a contradiction. In each case, one can avoid a particular contradiction by broadening the field, but only at the expense of introducing new contradictions. As long as one uses words, or "terms", with their associated thoughts, one is in the field of contradiction. For practical purposes, we can ignore these contradictions as long as they are below the "threshold" of our field of interest, but there is never any assurance as to how far it is safe to go on doing this. We need continual alertness and awareness to the whole process, if we are to avoid coming into conflict with the results of our own actions.

An important problem is then, "Why is it that in nature, there are isolable fields in which contradictions make no difference (i.e., they are below the threshold)?" This is a very deep question, but if we understood this, we could see what makes *knowledge* possible. For the general applicability of knowledge (which is always a conclusion drawn from past experience) to new situations, depends on the fact that the errors we make by thinking in fixed "terms" will, up to a point, make no difference. This will happen generally, if in the natural process, something is going on which is similar to what is going on in the human thought process. In other words, the existence of certain processes will elicit further functions, which in turn elicit functions of these functions, etc. All of this happening in such a way that in *the field of these functions* there is a certain stability and repetition. Thus, the "terming" process in our mind in some way reflects something analogous that (to a certain extent at least) also takes place in nature. Of course, the "terming" process is probably inadequate for a deep understanding of the subtler aspects of natural and human processes, and there creativity is necessary.

Finally, I would like to sharpen up the contradiction inherent in the *generality* of our concepts (see p. 7 of this letter). As I said, we perceive *difference*, and then we perceive identity as *non-difference*. But non-difference is perceived as *different* from any other kind of difference. Therefore non-difference must be perceived as the operation of the *difference function*, to register a difference from all other kinds of differences. Thus identity, which was first defined as *no kind of difference at all*, i.e., absolutely different from every kind of difference, is now seen to be a special kind of difference, viz., the difference from any other kind of difference. This is clearly a "counter-diction" and one which can in no way be avoided.

Likewise, the very use of the "difference" is a "counter-diction". Now any two things are by definition *different*. And such an example of *difference* is *different* from every other such example. Yet, we use the same word, thus implying that various examples of difference are not different. Clearly, in one and the same thought, viz., difference, we mean that each different is different, and yet, not different (i.e., the same) because after all, they are both examples of the same quality; viz., difference. Thus, we assert the identity (non-difference) of all kinds of differences, while we also assert their difference. At any one moment, both these meanings are present, although one may be uppermost in the mind. Logical thinking requires that both meanings shall always operate. And as I explained before, the same is true about the general and particular meanings of every word and thought.

Now, you might try to get out of this difficulty by saying that two separate examples of difference are different only in certain respects, while being the same in other respects. For example, the difference between an apple and an orange is different from the difference between a man and a monkey, because one is a difference in species of fruits and the other in species of animals. But they are both the same, in that each of them constitutes a *relationship of difference.* This will not however stand up, because you cannot be sure that the *respects* are identical with themselves. What do you mean by saying that yesterday's character of being a human being is the same as today's? Only that in the field of your mind at present you do not register a difference. On the other hand, it is clear that humanity (like anything else) has unlimited aspects and potentialities, so that literally speaking, it is a terrible and foolish mistake to ignore the possible difference between today's humanity and yesterday's. Neither the character of being a man, a monkey, an apple or an orange can be exhausted in the field of anyone's thought up to a given moment. Thus, we know that it is arbitrary to assert the self-identity of any quality. That is to say, it is a contradiction to assert that such a quality does not differ from itself. Hence, there is no way to compare two kinds of difference with regard to the *qualities* of the difference (except of course in some limited field in which nullity of difference can be defined). The whole way of thinking through comparison is nothing but "counter-diction". And perhaps it is not an accident that genuinely creative acts cannot be compared – they must be understood in their own frames of reference.

Very sincerely yours,
David Bohm

Red Wing
Route 2
Minnesota, USA
May 29, 1961

Dear Mr Bohm:

I'd like to begin with some questions. If terms function in the manner you describe, i.e., engender certain structural difficulties in the function of language that are serious frustrations, is it unavoidable that these terms must function as you describe them? If the procedure of beginning with "definite terms" leads to the linguistic conflicts you call contradictions and counter-dictions, are we doomed to begin in such a manner? What would happen, for instance, if we reversed your procedure? That is, instead of beginning with definite terms, begin with indefinite ones? *Undefined instead of defined terms*? In that case, all the *key* terms of our language – reality, democracy, science, art, socialism, atom, nature, quantum, etc. – would be, to begin with, undefined. Such terms would take on a defined structure *only* when a *particular individual* used them in a *particular context*. Such terms are not only capable of taking on an indefinite number of meanings but more specifically, as one writer put it, are multi-ordinal. This last makes it possible for language to take on an abstracting structure corresponding to the structure of present experience with the world. This attitude takes into consideration that the fundamental problems of language do not begin with language, but with pre-verbal abstractions.

Such a function of language is found not only in everyday usage, but even more so in such specific usages as that of physics. Difficulties arise, nevertheless, because the "verbal process," as it has been called, not only functions with its multi-ordinal terms, but this is largely unconscious (or inadequately understood). The consequence is the introduction (or retaining) of certain structures into language which are structurally destructive. Such is the notion of identity. Take the use of definite terms (in which the factor of identity invariably operates). What would happen if someone, say Pryce, was able to legislate that the term "science" constituted a certain definite term, and no other? I know you would agree that it would have a disastrous consequence for the further development of science. For science, like everything in the universe, is in a state of constant change. So our use of that term is likewise undergoing constant change. This is in line with your emphasis upon "thought process" as "constant movement." For this reason it is both futile and harmful to search for an answer to that perennially repeated question "What is science?" For it asks a reply in the singular (identity) about something that is many things, to many persons, e.g., your discussion with Pryce.

In dictatorial nations certain key terms are given apparent or disguised precise definitions by government fiat, which it behoves everyone to understand and live by, no matter what. These definitions contain taboos which dominate all fields, concerning such matters as relativity, psychoanalysis, art, etc. – anything that encourages the freedom to think as an individual. Even so, the demands of reality compel such forms of government to periodically alter the definitions (not to go into the kind of opportunist reality usage involved). In any case, there is always one official answer to such questions as, "What is art?" "What is science?", etc. To a lesser, but still pernicious degree, such verbal controls are practiced in democracies.

The "actual process of the world," you write, does not consist "solely" of words and thoughts. Could we make this sharper and so more accurate? Namely, the world *is not* words or thoughts. This has led one writer to remark, whatever we *say* anything is, it *is not*, for which condition you coined the expression "non-dictory" to distinguish the world from words.

In presenting the problem of the relationship between the function of the world and words, you end the first part by introducing problems of identity and differences. From certain things you say, I think you will agree if I put it this way: The fundamental relation between words and things is one of structure, with all the implications for abstracting which *order in structure* demands. This makes it possible for a structural correspondence to function between the non-verbal process within ourselves, and the verbal process, and the non-verbal process of the world. If this continuity of process is granted, which I know you do, then it seems that any notions of identity foster false-to-fact assumptions about process of any kind. The assumption here, one again you agree with, is that everything in every aspect of itself, is not only unique at every moment, but also in every one of its moments that have been and will be. Notions of identity state the contrary and involve one in contraries, what seem like kinds of "complementaries," as you argue the problem. The result is to frustrate efforts to bring the structure of the verbal process into correspondence with the actualities of the structural process within and without us. Differentiation, not identity, is the constant of experience. Identity engenders harmful either–or verbal fictions, introducing structures non-existent in process actualities anywhere. To deny this, it seems to me, is a denial of process.

In your theories, however, there is for me the curious situation of notions of identity along with equal emphasis upon the differential aspect of process. Curious, because old and new views are thus seeking resolution to problems arising from *their* meeting, by putting them into opposition to each other. It is as though the philosopher in you were carrying forward the identity notions of Aristotelian*ism*, while the physicist and mathematician in you were trying to maintain the non-Aristotelian views that pervade the latter two fields, by setting the two contrary metaphysics in opposition. This leads to your general orientation of resolving problems in terms of either–or counter- and contra-diction. Now I fully agree with your remarks on how we become unconscious of contradiction. But the question is, are the counter-dictions also unconscious contradictions?

Two objects have the "same" (identity) qualities, you say. Immediately, however, you qualify this by noting, that if these qualities were "precisely" the same (identity), "it would be impossible to assert their differences." You resolve this to me contra, not counter, diction by stating "they differ from nothing" under such conditions as you describe. While this is understandable, I would dispute this *method* of expressing the problem. Again, I would be inclined to reverse your statement – Everything differs by Everything. Nothing non-verbal can be said to "differ by nothing," since we know that everything differs by something from everything. It is better psycho-logics to state: I include so and so, everything else in the infinite process of the world I am conscious has been left out. In place of putting artificial stops to the process of our language function with "relative nothing," would it not be more process useful to say, at this moment I assert that between these two objects I tentatively assume certain *similarities*, not identity, of qualities, functions, or whatever? That would be my answer to your question, can I give another interpretation of "sameness"? We thus assume actuality resemblances (similarities), and avoid non-actual absolute sameness (identity). The door is then left ajar for the inevitable structural revisions. One is then sustaining a process orientation in one's general attitude. As you remarked, we must be constantly attentive to the fact that our reality experience is not the whole, i.e., we never know *all* about *anything*.

You both affirm and deny notions of identity. It would be "fruitless," you write, to use two different symbols on two sides of an equation, if both represented "absolutely the same thing." What then? Relative sameness? As with relative nothing? Again you say, we can use a term like "redness" in an "unlimited set of contexts," and yet mean the "same" (identity) thing. We cannot, you point out, use a different term for every different redness. Granted. But do such terms function as you describe? Do we not use some single symbol to represent, not the "same" thing, but an unlimited number of different things? We do not need a different term for every redness when the single term can symbolize to the infinity of redness differentiation, it being amenable to an unlimited number of contexts, as you would agree.

The aim, as I see it, is not to identify any experience of redness-yesterday with one today, or the latter from one moment to the next. I experience this very vividly in my work. When I first see one of my finished works, it is an incomparably sharper experience than takes place in either painting or sculpture. That is, each part is made in isolation and then, when all parts are finished, it is assembled all at once. Consequently, I see the finished work all at once! Now, what I *see* during that first moment, I will literally *never see again* in the *actuality* but only in my memory, where a residue of it will remain. For, from the first moment my seeing moves in a *constant process*, which movement is as relentless in its course as that from birth to death. Everything Is Changing All The Time, as even I am. Or rather, I try to facilitate this vision so that I will be changing in a non-frustrating way. One can say, the moment one sees the same (identity) work of art twice, one has ceased to *see* the second (identity) time. One is, so to speak, seeing largely with one's memory, projecting and so imposing the past upon the present. I might add, recalling some past vision of some particular work, that this activity plays an important role in my judgements concerning the progress of my vision, as well as the errors I have made along the way.

Now, you yourself state that it is erroneous "to fix the meaning of words" because they "never" possess the "same" meaning. Due, however, to your previous considerations, you are led to contend that each "definite assertion" *inevitably* produces contradiction. Such contradictions vanish if we discard notions of identity and regard terms as multi-ordinal, which is in accordance with actuality demands. Contradiction stems from the identification of the marks r-e-d-n-e-s-s as implying inescapable sameness, i.e., identifying words with the world. To confuse the world with the use of words, or, to make words correspond to the way we can use the world, that is the issue.

When we say, "I see," we mean, I understand. We don't say, "I hear." Vision, incomparably more than the auditory, is fundamental to comprehension of the world. If both sensory paths provide for our experiencing the world, vision is sharply unique. For it is a direct part of the brain. Accordingly, the most basic abstractions, in any field, are visual ones. We translate visual abstractions into the verbal process in a way that will achieve structural correspondence. This necessitates a *natural order* of abstracting, from the objective to the verbal process, not the reverse. (I realize this gets us up against some old arguments about the alleged autonomous state some attribute to mathematics, and which attitude I see as doing much harm in the new direction of art.) It follows that it is necessary to avoid confusing, specifically identifying, different orders of abstraction – words with things, the object with the "event" level, description with object, and so on with the ever higher orders of inferential abstraction.

In my *Evolution* (pp. 519–24) I consider the relation of mimetic art to the formation of the evolution of language. I conclude by considering the possibility that the new direction of art, a more purely visual form of art, will have a part to play in the future formation of language. That is, reintegrating the necessary and critical role of visualization in language, to correct the present over-verbalized interpretation of actuality.

The emphasis you give to the need for recognizing that, under certain circumstances, differences *appear* not to effect the particular observations of structure we are producing is, I agree, important. But I would not employ the expression "nothing" or "null difference," that being just what is done by those who, unlike yourself, are not aware of what they say, "This means nothing to me." I would prefer a term that makes a positive emphasis upon *remembering* the area beyond the threshold, which I think would more effectively accomplish what you emphasize, the need to be constantly aware of all the thresholds that constantly surround us on every side in everything we do. This term-psychology would more effectively neutralize the general tendency toward conservatism, i.e., to close off things, to imprison cogitation in conscious or unconscious absolutes. This tendency rests its destructive work precisely on the border of actual or imagined thresholds. Pryce offers a good example. He thought to build a concrete threshold around what constituted the practice of physics and philosophy. Your rejoinder made him, at least for that moment, a little more sensitive to the use of those two terms.

If nature is neither contradictory nor non-contradictory, but dictory as you say, are you not still faced with the question whether *you* experience nature as contradictory or not? I would say that nature is non-contradictory as I believe it is non-contingential, that such experiences only arise in human experience. What I have been trying to say does not deny contradiction, but rather rejects the inevitability of a verbal process which necessitates conflict and contradiction in its *inherent* structure.

You suggest that the truly creative act can produce an experience of totality, "transcending time." Is it perhaps that one is simply unconscious of time? The "deeper way" by which the creative act comprehends the world, to me seems to rest in its grounding in pre-verbal experience, and the capacity to return the results of the verbal process or thought, back to pre-verbal experience.

I am not clear why "genuine creative acts cannot be compared." Are there not diverse forms of human creation, all related but each distinct from the other, with informative similarities and differences?

In the creative act, as regards experiencing the totality, you might be interested in my own experience in observing nature, which I experience in art too. Initially, I try to experience nature on a non-verbal level of perception, maintaining verbal silence within myself. This achieves the following results. I put *aside* all I know, i.e., my consciousness of vision at the time, so that I will not function, at least with the intruding deliberateness of consciousness, with a *prejudiced* vision. This then permits me to submit myself as much as possible to the experience of nature, each time anew. At such times I have the experience of totality, not as knowing that totality of nature, but because I have removed the barriers to being *in* it.

You appear to have misunderstood me. I did not mean to imply the denial of mathematics to music. What I wish you to comment on is, why all the great composers, since the centuries after the research of the Pythagoreans, have not chosen to submit their composing to mathematics? This question seems particularly pertinent in the case of music, since it seems more amenable to the conditions of mathematics than the other arts. Just as important, why are those who do not compose music interested in applying mathematics to music? I am not aware that this question has been asked.

In our present discussions you have initiated issues basic to all our other discussions. Until now we have been talking about the structure of our interpretations of reality, now we are talking about the structural functions of our talking. This is the key to our understanding each other. Therefore, it would be a help to me toward that understanding if this time you directly answered my responses to your views.

Thank you very much for sending me the script of "Quanta and Reality." Although there were places I could not follow, I got enough of it to make very good reading.

> *Pryce*: (1) Physics is not philosophy.
> *Bohm*: (2) You have a specific philosophy.
> *Pryce*: I have a philosophy to avoid philosophy. (A devious attempt to salvage point 1.)
> *Bohn*: You always have some philosophy. (Sustaining point 2.)

Pryce gives a good example of using a definite closed symbol which in this case got him "in a trap" as he himself correctly put it. Your remarks on how to teach science theory to students is a breath of fresh air which has application to all forms of education. Your students are fortunate!

I look forward to your reply.
Best regards,
Charles Biederman

IV

Red Wing Jan 18/62

Dear Bohm:

(May we drop the Mr. stuff?) I am able to write to you much sooner than I had expected, due to a bad cold that does not permit me to work in my shop. I ignored the cold and now I'm paying for it. Last Monday your airmail, Jan 13th, and seamail, Dec 26th, both arrived. Here I will only reply to the Dec. 22nd letter.

Does the term "similar" really take us back to the very "samness" notions we wish to avoid? If "Everything is everything else," it can also be noted, from another position, that nothing is ever anything else, either as anything outside itself or anything it is itself from one moment to the next. We "see" what we call similarities because, as AK says, we leave out the differences due to the coarseness of our perception and/or knowledge. The further back we go in the evolution of man the greater the tendency to "leave out" the differences, and so the greater the reliance on identity logic. In the reverse direction there is an increasing sensitivity to differences, and eventually the notion of structure as process. Characteristic of this evolution is the greater relanance of experience to nature, with the lessening of dependence upon verbalisms. Absolute samness, absolute differences, neither exists, for the first would produce a frozen world of 'symmetry' structure, the latter a chaos of 'asymmetric" structure. But such symmetry and asymmetry do not exist in nature.

When we speak with AK of language as similar in structure to nature, we do not identify the two structures. We remember that words are never the things we speak about.

What is structure? you ask. Is there such a thing as structure? There are only structured things, and no thing that is structure. What do you say about that?

I could not have used the phrase "abstracting animal," as you have me say, nor does AK. Man is only an 'animal' when he acts something like one and then he becomes less than an animal. Man has the capacity animals do not, apparently no limit to his capacity to abstract. By time-binding Ak means that each generation can pick up where the last generation left off, and not simply repeat the last generation as animals must. This is not simply the quantitative distinction you make - perception-animals, abstraction-man. Rather, the distinction is quantitative-qualitative for both perception and abstraction, for both man and animal. How can a new stage of evolution go "beyond abstraction" by going "back" to a "new kind of concreteness" of totality perception. Perception itself is a form of abstraction, as are all our senses. It is these abstractions that are continued in the more familiar abstractions of the verbal process. Impossible to go "beyond abstraction" only beyond present forms of abstraction. And, any advance in abstraction entails a "new kind of concreteness,"else change is merely verbalism. Going "back" to a "new concreteness," a contradiction leading to pathology however "new." To be "new" is inescapable, choice is only, pathology or sanity.

There is a good deal more than "some sense" in your view of creation as "instantaneous thought-feeling-action." It is a culminating joy for the one doing the creating. Before (and after) the act of creation, one has done much preparation. And, is it not true that if this preparation is done with security demands sitting on one's back, the moment of creation will be 153

TRUTH AND UNDERSTANDING

3, Berkshire Road
Bristol 7
Glos., England
December 22, 1961

Dear Mr Biederman:

After a long delay, the time has come to continue our correspondence. I hope that you received my brief letter of a few weeks ago explaining the cause of the delay. Perhaps now things will settle down again, and I will have more time to answer your letters.

Let me begin by summarizing what seem to be the main recurrent themes of our correspondence.

1 The question of creative "termination" or "determination", as distinguished from either mechanistic determinism or pure chance and chaos. In this were certain subsidiary questions:
   (a) "Contingentation" and "necessitation". You admitted the category of contingency only for human beings, with their abstractions, but supposed that, in some sense, necessity applies to all of nature. I argued in favour of contingency as applying to all of nature too.
   (b) What is "order" and what is "disorder". Somehow you object to the notion of disorder, and have a similar objection to the notion of chance.

2 The problem of oppositions. You insist (quite rightly in my opinion) that opposites are never separated, always interwoven from the very beginning, and that they are only introduced to help us to analyse nature (not actually existing as such in nature). However, I maintained that there is something in nature that these opposites are reflecting, and you seemed to be denying this.

3 The problem of time. Each moment traces its past and projects its future in a unique way. At each moment there is an infinity of possibilities, which (we both agree) have, in some sense, a real existence. You argue that man can choose from these possibilities, and nature by itself cannot.

The past is not by itself fully determinate, but is ambiguous as to what it is (or was). The future helps to determine what the past really was. Thus, the past does not fully determine the future. (If only because in some measure, the future partly determines the past.)

4 The problem of totality. Originally, I defined totality as all that there is, was, and will be. You objected that this is too static. Rather, the totality is the process itself, acting creatively in every moment. Only the totality fully determines each part, and this totality is not accessible at any moment, to our knowledge. For the latter, by its very nature, refers only to the past, and not to the actual, living process, in its concrete existence from moment to moment.

This raises the question "What is a 'part'?" Are the parts introduced only by us for our own convenience, or do they in some sense actually exist? You agreed that the parts are contingent. In other words, the full reason for their being what they are is not contained in themselves, but only in the totality. If contingency is something referring only to our own perceptions and conceptions, then is the notion of "being a part" also only our own invention? Or is there some sense in which parts really exist?

5 The question of our relationship to the world. Observation without participation is impossible. We participate in what we observe and what we observe participates in us. Quantum theory shows that this must happen in physics, while common experience shows that the same holds in all our activities.

The separation of the observer from what he observes is then at best an abstraction of limited validity. It is impossible to avoid the conclusion that we are what we observe, and that what we observe is also ourselves. The distinction of "I" and "not-I" is only relatively true. But this distinction is at the basis of every distinction "A and not-A". That is to say, the relation between "A and not-A" is known to me, provided that I assume it is the same as that of "I and not-I". If this distinction of "I and not-I" is false, so must be that of "A and not-A". But as you state, process means (in the words of Leonardo) "Everything is everything else."

How do we understand the problem? It is related to totality. As I and not-I are in some sense, aspects of a totality, so must A and not-A be aspects of a total process. This requires a deeper understanding.

6 For the scientist and the artist, the question arises of how, as you say, he can "creatively deal with its past". He cannot ignore it or annihilate it. He must take it into account. Yet, he must not be dogmatically limited by past creative "decisions".

7 Questions connected with language, thought, and semantics (naming problems such as those considered by Korzybski). "The world is not words or thoughts." There is no such thing as identity. Fundamental terms (like "truth", "understanding", "totality", etc.) cannot be given a unique definition. Rather, as you say (with Korzybski), they are "multi-ordinal", or I would say, that they are implicit, behind all that we say and think, and yet not definable in an explicit way.

There is your suggestion that contradiction can be avoided, if we recognize that a term such as "redness" always has different meanings, never the same (i.e., it is multi-ordinal). Yet to me, this raises the problem of what we mean by "similarity". Our use of the same term "redness" surely symbolizes that in some sense, something is the same. If this were not so, why wouldn't we substitute another word for our later experiences of a given quality (i.e., roundness, blueness, etc., instead of redness)? I would say that this question needs a better understanding.

Your assertion that "seeing" and "understanding" are the same, seems worthy of more development. I agree with this wholeheartedly. We must go into the way in which we use words and thought, as you say.

8 We agree that, in art as well as in science, the key notion is process, with structure as the basic characteristic by which process can be described in abstractions. But what is structure? We know intuitively, but I think that a bit more reflection on this point would be useful. I would say tentatively that it is a kind of relationship in process, involving many characteristics (e.g., serial order of before and after, transitive order of inside and outside, questions of symmetry and asymmetry, repetition with transformation and fundamental

qualitative change, and many others). Have you further ideas on the subject?

9 There is the interesting notion which you gave that man is an abstracting animal (or as Korzybski would put it, a "time-binder"). This is no doubt at least an important part of the truth. But you yourself argue that man is undergoing further evolution, which must now be consciously understood, if the result is not to be disastrous. May we not say that animals work mainly on the level of immediate perception of concrete situations, while man works on the level of abstractions and time-binding? But perhaps the next stage of evolution is to go beyond abstractions and back to a new kind of concreteness, which is perception of totality. In this regard, science has something to contribute, and art a lot more, but what is most important is man's understanding of his own total process of existence.

Let me say a little about understanding. For example, one might begin by studying circles empirically, measuring the ratio of the diameters to the circumference, etc., tabulating the results in a log book. Then one might develop a geometry, defining a circle as a curve traced by a point moving equidistant from a fixed point, etc. The properties of circles would now follow from this geometry. Then, when the geometry was explained, you would at first follow the parts of the argument, but suddenly you would say "I see". By this you would mean that you now understand. In other words, you would no longer perceive the relationships in the circle by starting from the parts and trying to put them together, but rather you would start from a total process of thought, in which the circle, diameter, etc., would all be generated as aspects or sides, automatically in their correct relationships. Understanding comes in a "flash", or is felt by some people as a "click", in which everything falls into its proper place. Of course, when you perceive this totality, you do not see everything in detail as all at once. This comes out later only as you unfold and develop the full implications of such an understanding. But what is basic to understanding is that you suddenly cease to see certain things as parts that have to be put together, and instead see them as sides or aspects generated in a total process, so that you now understand why they are related as they actually are.

Of course, if you asked me for a more detailed definition of the word "understanding", I could go into a long story, but then, at the end, I would ask "Have you understood?", thus giving the game away. In other words, if a man did not at the very outset understand what is meant by understanding (even if he wouldn't put it into words), he would be worse than an absolute idiot, and no conversation with him would be possible. Similarly, every man understands what truth is, even when he is lying (consciously or unconsciously). Thus, it is no use trying to define such terms. But one can explain or explicate them (i.e., bring out explicitly a few of the relevant implications of these terms).

I think that everything important is really implicit in our thoughts and true communication (or conversation) consists in having two people create in each other trains of thought and feeling having essentially the same implicit content (i.e., each in effect, opens certain doors in the mind of the other, while he is doing the same for himself). (Here, we have the problem of what we mean by the word "same", but for the moment, let us leave this in the domain of the implicit.) Just as each note in a musical composition has no meaning, but the meaning is only in the composition as a totality, so the meaning is not in the separate words, but only in the totality of what is being said.

In this regard, I agree with you on another important question, that of whether thought and feeling can be separated. Of course, thought, feeling and action, are just three aspects of one total process. In anything fundamental, all three are inevitably involved. For example, can you say that truth is purely intellectual? Does one not feel that something is true, and is not this feeling just as necessary as the intellectual content that is thus felt to be true? Could there be truth without the feeling of truth? Could there be understanding without the feeling of understanding? Evidently, in everything deep and fundamental (involving totality), thought and feeling are just sides of a single process. Moreover, thought–feeling of this kind leads directly to an appropriate action. For example, suppose that you were reaching for a bottle that you thought contained sugar for your coffee, and suddenly you read on the label "Potassium Cyanide". Now, the label is of course a purely intellectual symbol for a certain substance. Nevertheless, the perception of this label would work in your feelings and actions. You would not need to debate with yourself, to think, to choose, etc., asking how this would affect your career, your bank account, what people would say, etc., etc. Immediately, you would perceive that all these questions are secondary, because the totality of your life is involved, and that if you took this substance, the other questions would not matter. Appropriate action (withdrawal from the bottle) would start as soon as the truth was perceived.

I think that in some sense, creativity is instantaneous thought–feeling–action. However, every time we perceive something, we generally call up a cloud of abstractions, and the consequent indecision in action. Of course, I don't want to argue against abstraction as such. For, as we agree, abstraction is one of the things that makes man what he is. Abstract conceptions may be compared to mirrors that distort reality, giving us a set of views that are magnified, demagnified, filtered, coloured, etc., etc. These abstractions are useful just because they distort. For by holding fixed that which is really changing, they enable us to examine certain relationships at our leisure. By leaving out certain things, they enable us to see others better, etc., etc. But our weakness is that we tend to get lost in these mirrors, mistaking what they reflect for the whole world. We must also perceive that our conceptions are only abstract reflections, and we must perceive the relationships by which they do reflect reality. I would say that we need perception of totality or understanding, which I symbolize by the letter P in the diagram below.

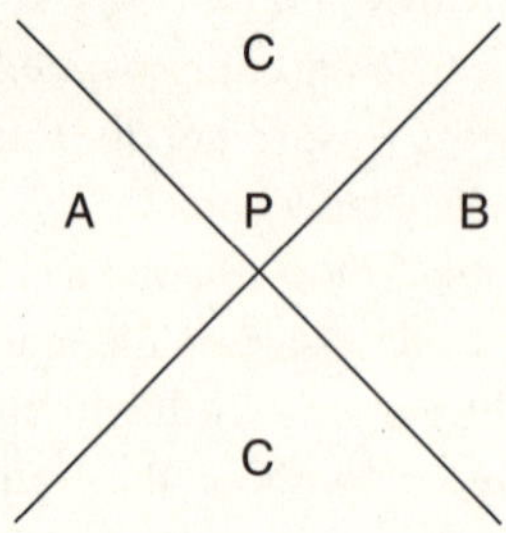

The perception of totality starts from a partial perception of the outer world, which I symbolize by A. Then, there is the perception of the inner world, containing our concepts, symbolized by B. (A and B considered separately are as far as most people ever get.) Then there is the perception of the relationship between A and B, symbolized by C in the diagram. Finally, there is the perception, P, of totality, in which A and B are seen as sides or aspects of a total process, automatically in their correct relationships, C. This total perception requires that there be no separation of "I" and "not-I". Each is only a side of a totality.

It seems to me that in all moments of creation (scientific, artistic or otherwise), there is no consciousness of an "I". There is only a process of creation. Later, when one reflects on the process, then one begins to think "It is I who did this thing. Isn't it marvellous what I can do, etc.,

etc." In other words, the "I" is like a chattering monkey that likes to take credit for everything, but actually only gets in the way by making a terrific lot of noise. When it is silent for a moment, then creativity can come into being.

So to come back to the new evolution of humanity, I would say that we have to go beyond our abstract conceptual way of thinking and come back to perception of the concrete as basic. But the concrete now contains the abstract, i.e., as a reflection, which helps throw light on the totality.

Truth is in an adequate understanding of each moment. In this sense, it is concrete. If I understand one moment, I may not understand the next, unless I remain alert and attentive to the total content of what is happening, because something basically new may enter meanwhile. There is no fixed, absolute, everlasting truth. Rather, it is a process that can come into being from moment to moment. It is neither subjective nor objective, but something that transcends the distinction between subject and object – and which therefore has to do with totality. Falsity is also a process, and we must understand the false as false, the truth as true, along with the true in the false. Generally speaking, falsity comes from mistaking a part for a totality. The truth in the false is to see this part as a side or aspect only.

It seems clear that falsity comes from using inadequate abstractions. Very often this comes about by our holding on too long to past abstractions, which were once at the very focus of the creative process but which no longer fit the new situation. In a highly technical society, such as ours, there is also an enormous number and variety of abstractions. We use them so often and so unthinkingly that we get lost in them, ceasing to live in a real world, and living in an imaginary abstract world instead. In other words, we forget that abstractions are abstract, and mistake them for the concrete living totality. Or it may be put in other terms that we allow conception to limit perception (i.e., we only see what fits into the past and known patterns of things). In reality, conception should serve perception (as a kind of mirror). Then there are those people who wish to insist on external perception only. This too is wrong, as we cannot perceive much without our conceptual mirrors. What we must perceive is totality, including the outside and the inside as sides or aspects, automatically in their proper relationships.

Now, I think the new phase of evolution of man is this immediate perception of concrete truth as a totality to be understood wholly and fully from moment to moment. This perception is neither intellectual nor emotional. Rather, it is thought–feeling, and gives rise to a total action, which is appropriate to the situation of the moment. It may properly be called creativity, but need not be restricted to art or science. Rather, it may occur in any moment of human relationships, as in relationships between a man and nature, or between a man and ideas. It contains abstraction. Indeed, each new moment calls for a new kind of abstraction. But it is, in a sense, a "non-abstract" kind of abstraction. For this abstraction does not stand by itself. Rather, it is perceived as being just what it actually is, i.e., an abstraction, and its relationship to the concrete totality from which it comes is also perceived. Thus, fundamentally it is a perception of a new kind. But in a sense, we have always known it, without recognizing it by name. For is not truth basically a perception too? And is not understanding also a perception? We use the word "see" for "understand" mainly because in vision we have the concrete perception of a totality, whereas in hearing, this perception is strung out in time, so that the totality is perceived only in the trace left by the whole process of hearing what is said. But understanding and truth go utterly beyond vision, hearing, feeling, or any of the particular senses. Even a blind man can "see", although he may never have had any vision from the time of his birth. In fact, I would go further and say basically that it is our understanding that is behind the impression that the world that we see is a totality. And when we understand the world as a totality, then for that moment at least, we are a totality, without the false division of I and not-I. Therefore, understanding is also a process, a total process, which is a possible mode of being for man. In this state of creative understanding, a man is an integrated whole, and not a set of varying fragmentary motives, desires, urges, conditionings, inner confusions, etc.

If you are going to ask what state of feeling goes with understanding, I am afraid that it will have to be described by the word "love". This word has unfortunately been used in so many false ways that it hardly means anything nowadays. Yet, I think that by implication, the meaning will come across. For example, some parents claim that they "love" their children, but do not understand them. Is this really possible? If they do not understand what their children actually are, then the beings for whom they feel love must be imaginary, just projections of the parents' own minds. Thus, what the parents actually "love" is not their actual children, but rather, some projections of themselves. Such a love is evidently false. Evidently, there can be no real love without understanding. Vice versa, can there be understanding without love? If we hate something, we reject it and do not understand it. (This is perhaps what disturbs me about your attitude to Picasso, Action Painters, Surrealists, etc. Not that I think that they are right, but one must somehow penetrate into why they say what they do, without being lost in their point of view.) If we are indifferent to something, we will never undertake the ardous task of understanding it. If something pleases us, we will be afraid to look at its dark side, and again we won't understand it, i.e., see it wholly and totally. So it seems that the only feeling that will lead to the action of understanding is love. Sometimes one sees parents, for example, who know both the good and bad sides of their children, and simply love them. Or sometimes, the artist or the scientist loves his work in a similar way. When this sort of thing happens, there is a total process, which is comprehension on the intellectual side and love on the side of feeling, so that, in a sense, the person is that which he understands (there being no separation of I and not-I). Such a state cannot be static. For under the motive power of these feelings, the person takes continual action, and also makes an effort to keep his understanding adequate to the situation as it develops. He is not doing what he does just for gratification of desires, nor out of anger, hate, jealousy, envy, or the need to feel secure (which are the usual motives for almost everything that we do). And out of these latter motives, true understanding cannot come.

I have gone into the question of understanding rather thoroughly because I think it is at the root of all the other questions. If I tried to go into all the questions indicated earlier in this letter, I would write a book. Eventually, I hope to be able to get to them. But my views have evolved a great deal since I last wrote to you, and it would be necessary for me to go into a lot of scientific material to say what I want to say about these questions. Probably I shall do this in another letter. Meanwhile, I shall make a few remarks on certain semantical questions, which are closely related to what I have just said.

In your letter of May 29, 1961, you objected to a suggestion of mine that identity means "differing by nothing". I admitted that it was not happily put, but now, I have what I regard as a better way of doing it. I begin with a concept of a field. Roughly speaking, this is what it means in common experience – that is – a domain which remains within itself as it moves in the context of certain processes. In mathematics, the field has a very simple definition. For example, the numbers form a field. For if you consider the operations of addition, subtraction, multiplication, or division of any pair of numbers, you get another number. Symbolically, if a and b are numbers, then a+b   a–b   a*b   a/b

are all numbers. Thus, the field is closed under certain operations. Its second important characteristic is that it has a zero as null element, (0), such that

a+(0)=a   a*(0)=(0)
a–(0)=a   a/(0) does not exist (that is, it is infinity)

I want to generalize this notion of a field. For example, we may call art a field, science another one, the field of vision, the field of our consciousness at a particular moment are still other ones, etc., etc. A field is something that moves into itself, reflecting and projecting the movements in some more general domain as it does so. Every field has a null element. The null element does not represent absolute nothingness, but merely those movements which cannot be reflected in the field in question. Every field is, in effect, defined implicitly by its null element.

It is clear then that two processes, A and B, are identical in a certain field, if they differ by the nullity of that field, i.e., by something which produces no effects in the field in question.

The idea of a field can be extended mathematically, especially with the aid of the concept of a group, but I needn't go into this here.

Now, let us come to the question of qualities, such as redness. It is clear that we know redness through a tremendous number of memory traces, tied together by a name (as some other symbol), in such a way that the symbol elicits certain sensations that we had when we actually experienced redness on earlier occasions, when the sensations elicit the corresponding symbol. Of course, actual redness that is immediately present before us is not the same as the memory traces by which we know redness and recognize it (any more than an actual dinner is the same as our memory of a dinner). However, in a limited field of certain sensations, the actual red differs in its effect from that of the memory traces by the nullity of the field. This means that in the totality, no two things can be identical. Yet, in certain limited fields, they can be identical (i.e., differ by the nullity of that field).

I think it is necessary to save the content of the concept of identity in this way, or else we could never justify the common usage of words. The notion of a field enables us to see what is the real content of this conception, without committing ourselves to the notion that there is actual identity of things that are different. In this way, moreover, I think that we can avoid the need for contradiction. For now, our words are recognized to refer, not to the process as it actually is, but rather to the trace of the process in a certain field. Our concepts are then only reflections, traces or shadows of reality, so that we are not surprised that we come to contradictions, if we say that they are reality. Reality is, however, actually only implicit. For it is that total process which leaves these traces, and does an infinity of unknown things besides. So perhaps we will be able to agree that contradiction is not necessary, if one understands the relationships between words, thoughts and actuality.

Now, all of this raises an interesting question indicated in your last letter. For since process is continuous, there must be a structural correspondence between its verbal and non-verbal aspects, such that in the verbal field, there can be a projection of the total process (i.e., a kind of shadow). But in order that this may be possible, there is a still more fundamental requirement: viz., the total process must be projectable. Now, in the mathematical theory of groups, one finds a natural "language" for raising such questions. For there are groups which operating on themselves just turn into themselves, and these groups contain sub-groups, which also move into themselves. The interesting point is that certain kinds of sub-groups stand in such a relation to the full group that the movements within the latter can be projected into the former. This is just the kind of relationship we are looking for. That is, we want the non-verbal domain to be projectable into the verbal field. Of course, I do not want to say that the world is really just a bunch of these groups. Rather, I suggest that these groups provide a good way of mentally and symbolically tracing the total process (better than the arbitrary everyday language) because these groups are in themselves totalities, and because they have in them sub-totalities, which trace the full totalities. Thus, the language of groups stands in better correspondence to the total process than does our everyday use of language.

I think that this point of view is more clear than the notion you propose of "similarity". For when you try to say just what similarity is, you are finally forced either to reduce it to an undefinable something, or else you say that two things that are similar are the same in some respect. For example, two triangles are similar in that they both have three sides (i.e., they have the same number of sides); they differ in the lengths of their corresponding sides. I think that similarity is too vague a concept to assume as an undefinable. Moreover, it seems to me, that two things with the same quality are in some respects the same, e.g., they elicit in me certain sensations which differ by the nullity of the field of sensations – which is to say – their difference is below the threshold of what is perceptible in this field.

There is a further insight here into the quantitative and the qualitative. The quantitative always requires a common measure, e.g., inches, apples, degrees, etc. Behind the statement that 1 metre=100 centimetres, there is the implicit assumption that all centimetres are the same. But each centimetre (like each apple) is different from the others. It is in a different place, built out of different piece of matter, etc. It would not, however, be adequate to adopt your suggestion of saying that all centimetres (like all apples) are similar. Rather, in a certain respect, they must be the same (or else counting makes no sense). In my point of view, this is simple. For all centimetres are the same in a certain field (for example, if they are placed side by side or on top of each other, or if they are projected on to the same line with the aid of light rays, telescopes, mirrors, etc.). From this point of view, it is clear that the quantitative comes from adopting a field in which qualitative differences are projected on to nullity. Therefore, the qualitative is logically prior to the quantitative, and the immeasurable to what is measured. The fact that quantitative change leads to qualitative change is then not mysterious. For the quantitative is an abstraction from the qualitative in the first place (a projection into a certain field). Therefore, we merely discover that this abstraction has a limited validity, so that when we push it too far, we are driven back into the qualitative again. For example, when we heat water (adding heat energy and increasing the temperature) at first there is only a quantitative change, but at 100°C, there is a change of quality – the liquid becomes a vapour. But in reality, it was always undergoing a series of qualitative changes, which were projected on to the nullity of a certain field (i.e., as to whether the substance behaves as a solid, liquid, or gas).

You can see also that the concept of nullity in a field is closely bound up with abstraction in general. Thus, it is not enough that we abstract from the concrete totality; for this abstraction would be of no use unless the totality were abstractable. For example, if we take a sheet of glass, we could shatter it into fragments, and discuss the fragments in abstraction, but we would no longer have a sheet of glass. The proper kind of abstraction is different, however, in that it does not imply fragmentation, but rather, a correct reflection or projection of the totality. This is possible because the totality contains sub-totalities into which its movements can actually be projected. Abstraction consists in considering one of these proper sub-totalities. But if we mistake something else for a proper sub-totality, then our abstractions will be erroneous, and will lead to contradiction and fragmentation. This seems to be what has been happening in a great deal of modern art. The old framework of abstraction has been burst, but many people are now abstracting according to an arbitrary principle of selection, thus producing a fragmentation, rather than seeking the natural principle that is implicit in the problem that they are presented with.

This leads us to a further point that you raised in your last letter; i.e., the need for a natural order of abstracting from the objective to the verbal (or artistic) process, not the reverse. You can see that we agree on the need for this natural order. However, I think that the concept of the nullity in a field helps give a clearer idea of what can be meant by such a natural order. I think also that it accomplishes what you suggest; i.e., we must always remember that there is an area "beyond the threshold". With group theory, for example, if one looks at the problem in this way, one obtains, for the first time, a mathematical language which, as it were, is always saying, in each breath "This is an abstraction, and everything in the nullity of the field is projected into nothing."

The basic question is, as you say, whether we must experience the world as contradictory. Probably you are right in claiming that this is not necessary, and that a purely visual (non-mimetic) art will help in the evolution of a better way of using language. (For example, since vision suggests a totality, this may help us to think and speak with a totality always in the background, present by implication, so to speak.) Mathematics, physics, etc., may also help, along the lines indicated in this letter. However, what is most essential is obviously understanding, and this is where the chief problem lies for all of us. Here, we must be careful to pay attention to our own ideas, tracing their roots, and seeing where they lead. In this connection, our feelings play a key role. Are we approaching the subject with love, or with hatred, anger, resentment, envy, irritation, annoyance, etc.? The latter are a basic source of contradiction in our thought process. Of course, all of us feel that we have plenty of justification for some of these feelings (and the fact is that we probably do have it), but we will have to ask ourselves what is more important, to justify ourselves, or to understand truth as it actually is? I know that I often feel such irritations against most physicists, and I can see that at times, you have such reactions against Picasso, *et al.* But for both of us, the question is "Is it worth it?"

With best regards,
David Bohm

P.S. I think you can see why certain feelings tend to lead to contradiction. For example, suppose your main interest is to gratify certain desires (for power, money, fame, sexual satisfaction, or what have you). Then you are not interested in nature or other people as they actually are. You are interested in them mainly in their possible role of gratifying your desires. You do not bother to understand the total process, but are satisfied to find techniques that give you the gratifications that you seek. You don't worry about side effects of these techniques, which create new problems, but hope that further techniques can be found to deal with these.

Evidently your mind is in a state of contradiction at the very outset. For you do not want to accept what actually is – if you did – there would be no desire to be gratified. Nor do you want to understand what actually is. All that you want is to have a picture of the world which gives you the pleasant assurance that your desires will be gratified now and for ever, if possible. This means that the mind must reject what actually is, and set up an imaginary state that seems more pleasant, towards which it tries to move, with the aid of various techniques, etc. Immediately, you are in a state of contradiction, because one part of you knows (at least vaguely) what actually is, and the other part of you is struggling against it. But you cannot carry out a struggle externally unless you carry out a corresponding internal struggle – and this is the contradiction. For example, you cannot be angry at someone without being angry at the image of this person within yourself. This anger is directed within, as people discover when they get ulcers, high blood pressure, and heart failure from persistent anger, and aggressiveness in their ways of life. The energy of aggressiveness is directed against the image within, but the image cannot "take it". The energy goes on to work destruction in the body. But it also wreaks havoc in the mind, creating tension, fragmentation, confusion, and a general state of internal contradiction, which is the inevitable concomitant of having one part of the mind (i.e., "I") fighting another part (i.e., the image of the other man), when in reality, both parts are basically one and inseparable.

In other words, any emotion that leads a person to reject and struggle against something external will lead him to do the same against a part of himself. He will therefore be in a state of internal contradiction, in which understanding (perception of totality) is impossible, since by definition, he is limiting his perception to some part of the total internal field, with which part he identifies himself. Thus, there is initiated a process of disintegration and confusion, which tends to spread, causing further and further fragmentation of the individual. On the other hand, if he understands the totality of his field, he cannot be rejecting anything. This doesn't mean he is accepting it either. He is simply perceiving the totality as a totality, i.e., such that each part, including himself and what is not himself, are sides, generated in their proper relationships in the total process. The corresponding state of feeling is what is usually described by the word "love". He is then actually an integrated person, without fragmentation, whole, etc. In such a state, he may understand the whole situation deeply enough to take the appropriate action, without introducing a self-contradictory state of internal resistance of one part against another within himself.

So you see I think that there is a very fundamental connection between thought and feeling.

One more point. In your last letter, you said that you didn't understand why I maintained that creative acts were not really comparable. You felt that they could be compared. Perhaps I did not make myself clear enough. I meant that in the essential features which are what make a given act creative, there can be no comparison. Of course, they can always be compared in many respects, but I claim that in this way, one will miss what is most fundamental to creativity.

If you keep in mind the fact that each creative abstraction is in some way new, and appropriate to the unique and peculiar total situation to which it is a response, then you will see that in that regard, it cannot be compared to another creative act. For to compare two things, you must put them into the same general framework of conceptual abstraction (i.e., the same field). For example, you can say that A is larger than B, but the field of size is present before you have A or B, so that the general notion of being larger in size can hardly be regarded as a creative abstraction (i.e., if B is already greater than something else, C, then A does not introduce something new and creative just by being greater than B, since the relation of being greater than is already an old and known relation). I think you will see that all comparisons are of this nature; they must work in some field that is common to the two or more things being compared. Therefore, they must leave out something else, and what they leave out is just the feature of unique creativity, which can be understood only in the special field of the totality of that act of creation. In other words, I claim that a truly creative act, to some extent at least, creates its own abstractive frame, and for this reason, it cannot be compared fully to anything else.

3 Berkshire Road
Bristol 7
Glos., England
December 26, 1961

Dear Mr Biederman:

This letter is a supplement to one that I just wrote to you. It is intended to say a bit about my reactions to Structurist art. (You are probably saying "At last, and about time, too!")

First of all, I haven't managed to see a great deal of it yet. I have been busy with other things. Of that which I have seen, Carl Visser impresses me as the best. I haven't seen any of your work, except in photographs, but some of the photographs suggest a very good use of colour. The spatial distribution of surfaces is hard to judge from a photograph, but some of them look as if they would be very interesting to see.

Secondly, exhibitions are, as you have remarked, a bad place to show Structurist works. Logically speaking, each Structurist work should stand in proper surroundings, with which it tends to integrate itself. A flat wall of a museum, surrounded by other works, is not the place where this kind of integration is easily accomplished.

Thirdly, a great deal of the work that I have seen strikes me as cold, calculated, lacking in the kind of freedom that is the essence of true creativity. Perhaps it is too mathematical, and not irregular enough. In this regard, Anthony Hill once made a revealing remark to me that he feels himself to be a kind of frustrated mathematician. There is perhaps a tendency for this work to attract the mathematically minded. In this regard, however, some of the photographs of your work suggest warm glowing colours and an inner life, which goes beyond mere mathematical calculations.

May I ask you a question here? Is it necessary for Structurist art to restrict itself to simple planes? Doesn't nature structure have curves in it too? And aren't there rich and complex gradations of colour in nature? This would suggest that there may be a place for something like painting after all. Do you have to restrict yourself to sheets of plastic, metal and glass, with their extremely simplified and antiseptic appearance? Perhaps this is part of the reason why the net result tends to come out very mathematical and calculated, at least when many people try to put these principles into operation.

I think that I understand in part your reasons for working with these very simplified structures. Perhaps you are trying to start with something easy and then to develop your understanding further, before you go on to curved structures and complex colours which, as yet, you do not know how to integrate into Structurist art. In other words, perhaps you are still in a very experimental phase, and in a preliminary part of the experiment at that.

There is another reason for favouring planes and lines, which is that there is some evidence for the underlying discreteness and discontinuity of natural process. Physics suggests this in the quantum theory, which shows that all process is discrete and indivisible, involving "quantum jumps" at the micro-level. In the macroscopic domain, a typical process has so many of these jumps that it appears continuous (like a rain of grains of sand), but underneath, it is discrete and atomic. In a similar way, the Impressionist breaks up complex colours into a large number of pure colours, each discrete and sharply defined. The Cubists wanted to represent nature by discrete planes. And you are going on from Cubism in a similar way to a three-dimensional non-mimetic art; but continuing these particular basic principles of Cubism? Do I understand you correctly in this regard.

So perhaps you are developing a new kind of "creative abstraction" in the form of planes in three dimensions, with special emphasis on relief structures. If so, then several questions arise. First, are these abstractions "natural", or are they being arbitrarily imposed by your own personal peculiar background of thinking, conditioning, habits, preferences, urges, etc., etc.? We both agree that Picasso, the Expressionists, the Action Painters, etc., tend to fall into the latter category. But what about Structurism? I realize that with something new like this, you need a lot of experience before you understand it well enough before you understand its truth or falsity. You have to "listen", as it were, and to pay attention without judgement, but just with the desire to understand. Intellectually, I find what you say convincing and attractive, especially because it ties up with so much in physics and mathematics. But it still does not move me in my feelings. This doesn't mean too much from your point of view, but you might be interested in the way the whole thing has struck me.

In this connection, your analysis from "M. to M." is very good. However, there is always a danger that you may be carried away by the brilliance of your own analysis. Certain trends in art appeal to you in the first place. Perhaps this is a sign of something deep, and true about them, and perhaps it is a sign that your own conditioning leads you to favour certain things. Then you engage in a very good analysis, which reinforces the trends that appealed to you in the first place. Do you see the danger in this? Not that I am suggesting that you are wrong. At present, I am hardly competent to do this. I only wonder if at times you don't throw up the whole business temporarily and try to see it from another point of view? Do you ever get very sick of everything that you have done, feeling that it may be false, superficial, inadequate? I sometimes feel that we must do this from time to time, to avoid getting stuck in a groove. (At least, I do this in my work from time to time.) Discontent is necessary, or else one will go to sleep.

Another question that arises is whether the Structurist type of abstraction isn't just part of what is necessary. As I tried to explain in my previous letter, I think each new moment calls for its own unique type of abstraction. Is the arrangement of coloured planes in a relief a rich enough form to accomplish what is required? Evidently you feel that it is, or else you wouldn't work in this way. Perhaps you could go in more detail into why you chose this form?

Thirdly, there is a problem of getting theoretical ideas across, which is common to all fields. Evidently, you have had a vision of truth, seeing nature in a new light. Then you analysed the field of art in the light of this vision, and published the results. But then comes the problem. Young people come and read what you wrote, without being able to have the same vision of truth. In reality, nobody can see by anybody else's light. He must find his own light in himself. I know that you understand this and agree with it. What you write may be helpful to someone else, but only if he is engaged in creative work himself. The difficulty is that people tend to read what you write, and follow it without a deep understanding. No doubt, under present circumstances, this is inevitable, and there isn't much that can be done about it. It is just a fact. But it may be helpful to be aware of this fact, that many people are just looking for a lead, because there is an emptiness within them that they don't know how to fill. It is no use getting angry at them. They don't understand their own problem, and no doubt do the best they can within the framework that they do understand. What is needed when you write your analysis is also an awareness of the problem of trying to kindle the creative flame in other people (or helping them to see what is in the way). The way things are now, one is liable just to start another school, and the essential point of your own vision may be lost or confused. Perhaps this is why you have withdrawn to such an isolated place as Red Wing.

I hope that you are not offended by some of these remarks. The same thing happens in physics too. Once a new idea is hit upon by somebody, if it is successful, it is soon worked to death by a horde of people anxious to get their reputations, get good jobs, get ahead, etc.

Now, I would like to discuss my experience with the work of Carl Visser. We first saw this in a museum in Amsterdam, and recognized it as inspired by Structurist ideas. Later, we talked to Visser and had a very interesting discussion.

Among his works were several which appealed to me very much, particularly an earlier partly representational one of a bird in rushes. It was made of thin strips of black metal on a wooden base. There is no doubt that Visser was already coming to Structurism in this work, but he hadn't made the full change yet.

The fascinating thing about this piece of work is the development of inexhaustible new aspects with every change of point of view. Each time you moved, you saw something unexpected. Yet, later, you saw that this unexpected combined with what you already knew to make a single pattern, with room in it, however, for still more unexpected things when you moved again to still another viewpoint. There was a character of infinite, unbounded life, in this piece of work, and yet it was entirely in a certain very limited form, involving nothing but black strips of metal.

I can now see that Visser had captured this quality of contingentation and necessitation that we discussed in earlier letters. Each pattern was just incomplete enough to leave room for contingency, and yet this very contingency became the seed of a new pattern of necessity that encompassed the earlier pattern, while still leaving room for something fundamentally new.

While the element of mimesis was very weak, it was still there, and seemed to play a big part, for one got emotional associations of a lonely marsh under the autumn sky, with the solitary bird, all alone in this world. Somehow, I got the impression that these feelings played a big part in helping the work come alive, and that without this small trace of mimesis, it might have seemed "dead" and mathematical. This raises the interesting point of whether it is really possible to do without mimesis altogether. Slavish imitation of nature as we see it is not enough. We must play our creative role, bringing out possibilities that only we could have brought out. Yet, can we ever liberate ourselves 100 per cent from mimesis? And if we could, would it be right for us to do so? Perhaps the effort to be 100 per cent free of mimesis conditions us, and determines us as anti-mimetics. Just as the anti-Communist is determined by the very Communism that he opposes, so the anti-mimetic may be limited and restricted artificially by too energetically striving to avoid mimesis. Is it not better to say that the question of mimesis or non-mimesis is not the central and essential one (just as Communism versus anti-Communism is a false question – a red herring that prevents us from seeing what our problems really are)? In art, the essential and relevant question is, as you say, creativity. In being creative, we must free ourselves from the traditional duty of mimesis, but we do not need to define ourselves as anti-mimetics. At least, that is the way it seems to me. I would certainly appreciate your views on this question.

In some of Visser's first fully Structurist works, I felt that he lost this kind of free spontaneous character of life. He seemed to be struggling with something, probably the problem of order and disorder (which he said in our conversation was very important to him). As his work progressed, one saw a change from heavy symmetrical patterns to lighter patterns, in which an element of asymmetry was consciously and deliberately introduced. But one still felt that it was all too "planned" and not really as alive as this remarkable piece of earlier work.

A few months ago, we received some photographs of later works by Visser, in which (without mimesis) he seems to be coming closer to the style of the bird in the rushes. Perhaps he will finally work this problem out for himself.

This brings me to the problem of order and disorder, causality and randomness, law and chaos, on which we have touched many times in earlier letters. We have had some disagreement as to what these are, as to whether randomness and contingency really exist in nature (and not just in our own minds).

I have had some further ideas on this question since I last wrote you about them. Let me begin with the problem of throwing coins. In each throw, two results are possible, heads and tails (to be symbolized by H and T respectively). In a series of throws, we may get, for example, a very regular result, such as

H H H H H H H H H H H H H H

or

H T H T H T H T H T H T H T H T

But we can also get a result which we call irregular, such as

H T H H T H H H T H T T T H T

Now, it turns out that in almost all such series of throws the results are what we would call irregular, and only very rarely are they regular.

Now, what do we mean by an irregular or random series of throws? If we tried to say that something is disordered, irregular or random in itself, we should come to a self-contradiction. For no matter how complicated it looks, every order of results is some order, and cannot be called a "dis-order". But randomness or disorder can exist in a relationship between two or more orders. Thus, in relationship to the time order of throwing, most of the results are random (also in relation to the space-order of the places where the coins fall). By this, we mean that if we pick any time sequence (or space arrangement) containing a large number of throws, we will find:

1 that there are almost as many heads as tails in it (i.e., no tendency to prefer heads over tails);
2 in any sub-sequence, heads are followed by tails almost as often as by heads;
3 in any sub-sequence, there is statistically speaking, no correlation between the results of a given throw and those that are one, two, three, four, etc., throws later. (For example, if you look at all the cases that are heads, you will find that five throws later, there are tails just about as often as there are heads.)

If you reflect a while, you will see that relative randomness has a clear and well-defined meaning, with real consequences. For example, in a really random game, it is no use looking for a special "system" to beat the game, as this would imply a relationship between one order (e.g., time) and another (e.g., the results of throws) that does not actually exist.

In a really random relationship, you find what may be called an independence of two orders. In such a case, what is found in one order has statistically no correlation to what will be found in the other order. In the case of coin throws, the reason for this independence is clear. For in each throw, the result depends on a multitude of factors (e.g., just how the coin is held, with what impetus it is thrown, etc.) which vary from instance to instance in a manner having, on the average, no particular relationship to the time order of throwing.

The notion of independence of orders is very important. For such an independence leaves room for contingency. For example, as you follow one order (e.g., time), there is interwoven in it an unrelated order (i.e., the results of throws). By just giving the time, you do not determine the results of the throws, and by just giving the results of throws, you do not determine time.

More generally, contingency is possible, in part at least, because there do exist unrelated orders. In this sense, contingency has its roots in nature structure, and not just in our own minds. In other words, just as abstractions made by us are justified because nature is abstractable (see my previous letter), so our ideas on contingency are justified because nature is contingentizable. In other words, as we give one order, we can still leave room for the possibility of an independent order, not determined by the first one. This doesn't mean that nothing determines the second order. It only means that the second order is not determined by the first (and vice versa).

There is also of course the idea of related orders. For example, in the movement of the planets, the order of time is very closely related to the order of the series of positions occupied by a planet, so closely that this motion is precisely predictable, in the sense that if one reads the time on a clock, one can say where the planet is, and if one sees where the planet is, one can say what the time is.

Of course, there is every possible case between a perfectly defined relationship of two orders and perfect independence (or relative randomness) of these two orders.

Now, another important question that arises in the relationship of orders is the question of their compatibility. For example, consider a crystal, which is a perfectly periodic, repetitious, regular array of molecules. All real crystals have dislocations, which are breaks in the regular crystalline orders. Now, these dislocations are not absolute disorders. Rather, they are just another kind of order, which is not compatible with the crystalline order. The more dislocation in a particular specimen, the less will be the aspect of crystalline orders, and eventually if there are enough dislocations, the crystal order disappears. Here, we see how the development of one kind of order (dislocations) must be at the expense of another kind (the crystal).

Another such case arises with the molecular theory of gases. An ordinary gas consists of a mass of molecules in random movement. By this, one means that in relation to time, space, and large-scale order, there is, statistically speaking, no particular or distinguished direction or magnitude of molecular velocities, no distinguished position, etc. In other words, the two orders (micro and macro) are nearly independent, and this independence (with its relative randomness) is necessary for the regular behaviour of the gas on the large-scale level. For example, only if the micro-movements are random will the pressure of the gas be uniform and regular (as a random rain of grains of sand produces a nearly uniform pressure). If all the molecules were moving regularly and nearly in the same direction, however, they would start to collide, and as a result, there would appear violent turbulence and irregularity on the large scale, which would die out only after relative randomness between the levels had been achieved.

Here, we have another case when two orders are incompatible. Macroscopic order requires microscopic disorder.

When two kinds of order are incompatible, we can analyse the problem with the aid of certain abstractions. For example, in the case of a crystal, we can think of an ideal crystal without dislocations. We compare the actual crystal with the ideal crystal. If there are a very large number of dislocations, the actual specimen will become random in relationship to the ideal.

Here, we see how order is treated abstractly. An actual thing, with a mixture of two or more incompatible orders, is mapped into our ideal order. One will see that the actual order does not fit the ideal order. Every possible case can occur, from one approaching perfect fitting (implying a close relationship of the actual and the ideal orders) to no fitting at all (implying independence or a random relationship between these orders).

The problem of symmetry is closely related to that of order. Thus, could we not regard asymmetry as a kind of "tension" between two incompatible kinds of order? For example, an equilateral triangle has a certain kind of symmetrical order. But then there could be an asymmetrical order (e.g., one side is twice the second, which is in turn twice the third). This second kind of order is evidently not compatible with the first. However, one could try to make a pattern of triangles, which somehow interweave these two patterns of order. I haven't thought clearly how to do it yet, but a moving and developing pattern should, it seems to me, have this tension between incompatible orders. However, the tension should not be so great as to disrupt all relationship between them, and lead to relative independence or randomness. Yet, it must be great enough to break the rigid frame of each particular order, and to create room for something which is new and unexpected, not determined by the previous pattern, and yet compatible with it, and even more, capable of including it or comprehending it. In other words, there can be a tension which determined contingentation of the old, and subsequent necessitation of something new, necessitated in this way.

It is my opinion that in a moment of creativity, there is a "flash" or a "click", very swift, in which a new totality is determined (as in the case of understanding that I discussed in my previous letter). This new totality forms all its parts as sides or aspects, automatically in their right relationship. When such a totality is formed, it will be "alive", containing the kind of "tension" that allows new developments, which do not burst its framework. The thing cannot be planed or calculated by a mathematical formula, any more than one can plan or calculate the process by which one understands mathematical demonstration. (Recall the case I discussed in my previous letter of the sudden "flash" in which one "sees" the relationships of circle to diameter, etc., when a mathematical theorem is explained.) As we look at a work of art, we must come to understand it as a totality. (This doesn't mean all at once, of course.) In a complex and subtle work of art (also in mathematics) there will be many levels of understanding. The first step will lead us on to the first case of contingency, then we will stop and understand the next stage, showing the necessity of this contingency, in the framework of a new concept of the totality, and so on. A real work of art resembles life itself in being inexhaustible, so that one must understand it from moment to moment, from one occasion to another, bringing in all previous experience on each new occasion, so that everything is once again new. But this will happen only if, objectively, the work of art has contingentation, due to tensions between orders, leading to breaks in orders, that are understood in terms of newer and larger patterns, etc.

I think these are my main ideas on Structurist art thus far. Of the English members of this school, I feel that Mary Martin is one of the best. However, I feel that on the whole, the chief stumbling block is the question of contingentation and necessitation, and the "inner tension", needed to make the works "come alive" and not be an illustration of mathematical relationships.

I should very much appreciate hearing your reactions to these comments.

With best wishes,
Very sincerely yours,
David Bohm

P.S. With regard to the relationship of mathematics to music, which we discussed in earlier letters, the essential point is, perhaps, that the effort to present a mathematically calculated music would be an absurdity. The pure creative act, even in mathematics itself, cannot be calculated. What mathematician would undertake to calculate the process by which a new creative insight and understanding develops in him? There are a great many mathematical relationships in music. And indeed, there are in art too. For there is symmetry, group theory, and topology (the latter not having been exploited very much as yet, from the point of view of pure mathematical theory). Indeed, there is mathematics in everything. But from this, it doesn't follow that everything is just mathematics. Least of all can the creative act itself be reduced to mathematics. To create a new musical theme, a new work of art, or a new form of mathematics, one has to be in a living process. In this process, there must be feeling – passion – as well as intellectual thinking and analysis. For this feeling not only provides the motive power for understanding (as I indicated in my previous letter). It is also an integral part of this understanding. There is no understanding without the feeling of understanding, of truth, of totality, and this feeling is essential to the expression of the content of what is understood. What comes out of calculation is only the result of a technique. It is like pushing a button in one of these modern machines and getting coffee, doughnuts, chocolate, chewing gum, and all sorts of things. But when you understand something, then inwardly, you are it, and feel yourself to be it. Rather, you no longer perceive or feel a separation between yourself and that which you understand. On the other hand, the technician, the calculator, is always aware of a separation, a gap, between himself and the subject of his technical procedures and calculations. He engages in operations on a world conceived to be purely external to himself, in order to gratify his desire to exploit this world, to obtain results from it which satisfy him, give him a sense of power, money, recognition, or what have you.

Of course, there is nothing intrinsically wrong with technique and calculations. What is wrong is that they have become the essence, the totality, whereas they are meant to be secondary, and to serve the creative process and the understanding of totality.

Red Wing
Route 2
Minnesota, USA
January 18, 1962

Dear Bohm:

(May we drop the Mr stuff?) I am able to write to you much sooner than I had expected, due to a bad cold that does not permit me to work in my shop. I ignored the cold and now I'm paying for it. Last Monday your airmail, Jan. 13, and seamail, Dec. 26, both arrived. Here I will only reply to the Dec. 22 letter.

Does the term "similar" really take us back to the very "sameness" notions we wish to avoid? If "everything is everything else," it can also be noted, from another position, that nothing is ever anything else, either as anything outside itself or anything it is itself from one moment to the next. We "see" what we call similarities because, as AK [Alfred Korzybski] says, we leave out the differences due to the coarseness of our perception and/or knowledge. The further back we go in the evolution of man, the greater the tendency to "leave out" the differences, and so the greater the reliance on identity logic. In the reverse direction there is an increasing sensitivity to differences, and eventually the notion of structure as process. Characteristic of this evolution is the greater relevance of experience to nature, with the lessening of dependence upon verbalisms. Absolute sameness, absolute differences, neither exists, for the first would produce a frozen world of "symmetry" structure, the latter a chaos of "asymmetric" structure. But such symmetry and asymmetry do not exist in nature.

When we speak with AK of language as similar in structure to nature, we do not identify the two structures. We remember that words are never the things we speak about.

"What is structure?" you ask. Is there such a thing as structure? There are only structured things, and no thing that is structure. What do you say about that?

I could not have used the phrase "abstracting animal," as you have me say, nor does AK. Man is only an "animal" when he acts something like one and then he becomes less than an animal. Man has the capacity animals do not, apparently no limit to his capacity to abstract. By time-binding AK means that each generation can pick up where the last generation left off, and not simply repeat the last generation as animals must. This is not simply the quantitative distinction you make – perception–animals, abstraction–man. Rather, the distinction is quantitative–qualitative for both perception and abstraction, for both man and animal. How can a new stage of evolution go "beyond abstraction" by going "back" to a "new kind of concreteness" of totality perception? Perception itself is a form of abstraction, as are all our senses. It is these abstractions that are continued in the more familiar abstractions of the verbal process. Impossible to go "beyond abstraction," only beyond present forms of abstraction. And, any advance in abstraction entails a "new kind of concreteness," else change is merely verbalism. Going "back" to a "new concreteness," a contradiction leading to pathology however "new." To be "new" is inescapable, choice is only pathology or sanity.

There is a good deal more than "some sense" in your view of creation as "instantaneous thought–feeling–action." It is a culminating joy for the one doing the creating. Before (and after) the act of creation, one has done much preparation. And, is it not true that if this preparation is done with security demands sitting on one's back, the moment of creation will be nullified? At this point it is tempting to compliment oneself by saying it is necessary to be daring, but not foolish. But I think it more exact to say it is more a question of being natural in preparing for and then grasping the moment of creation. I concur completely with your notion that *all* forms of creation are characterized by the conscious absence of the "I."

True conversation, you say, is two people creating similar, I would not say "same," trains of thought to and with each other and in themselves. But this is nullified by our culture of too competitive "I"s. In its deepest sense conversation is a form of creation which withers before the conscious "I." "Same" conversation is like producing a leaf of absolute bilateral symmetry – identification – frozen symmetry. We meet not to repeat, but to create.

Yes, true understanding is a feeling-experience. But, why not look both ways? As I tried to show in my "Instinct . . . " article, there is a thinking–feeling and feeling–thinking spectrum. This implies the even wider spectrum of consciousness–unconsciousness. General tendency is to isolate the poles, leaving us without the spectrum between. Hence the obsession to decry the unconscious as solely the abode of Satan.

Your attitude towards abstraction impresses me as pessimistic, even though I agree with what you say. Abstractions are not as ephemeral as you make out. Certain abstractions, from the very earliest times to the present, remain as truths within *their* context. Even the genetic notion of the earth's structure as a flat surface remains a "truth" of *actual* experience, which you and I abide by daily in order to secure certain types of necessary experiences or actions. Not only do certain of the most early abstractions remain valid as such, but also because they are the departure for more sophisticated abstractions about nature structure. This works in the same way that the past remains a base for action with the future. These different kinds of orders of truth, originating in different historical orders, are possible because nature itself is a structure of differing orders of truths. Nature is so rich in its diversity that even in the case of any one order of these truths, there erupts again a plurality of truths. Abstractions "distort," you say. But isn't this only one type of abstraction? We can also distort abstractions! That is, stretch the truth of an abstraction beyond its reality context. "By leaving out certain things," you write, "they enable us to see others better, etc., etc." Just the point, only so do we see certain things better. If we had to include everything in order to see anything better, human life would be a floating dream devoid of emotional–intellectual depth.

When you refer to totality as various sided, this seems to fit in with the above notion of nature as various sided truth. If so it does not seem correct to say that understanding of the totality–truth goes beyond "any of the particular senses," that basically it is understanding that allows for the realization of totality–nature. Here I suspect that the physicist has indeed taken over who, after all, spends so much of his experience within the silent, invisible labyrinths of nature's atomic underground, where understanding gains a transcendental status over the senses. I suggest that the physicist is in a kind of sensory dilemma, *as a human being*. He must picture himself in a world devoid of human habitation, and where he is forced to rely too much upon understanding in terms of language. This is an inhuman situation and can only be relieved by relating this aspect of nature back to the human aspect of nature. Can this human dilemma explain the effort of earlier physicists to wave macro-nature into oblivion, and then the later effort to deny all picturization of nature as reality?

Anyway, if the totality is many-sided, perhaps many avenues of approach to it also exist, as the body has a number of sensory avenues leading to external nature. This is to suggest that understanding is a part of each of the many approaches to truth. A blind man can say "I see" to mean he understands. But, have you any hopes that there will ever be a blind Poincaré, Rembrandt, Beethoven?

I agree unreservedly that anger, hate, jealousy, envy, need for security are gratifications that prevent us from seeing the truth. Love can also, however, serve similar gratifications. As in your example, parents who "love" their children yet do not *understand* them because they love only a self-projection. Love like hate falls prey to ignorance of truth. But those parents who *know* the good and bad about their children, you say, can "simply love them." In each example you had to include more than "simply love" – love of the experience and knowledge of actuality or disregard for it. This last brings me closer to my view. Earlier in your letter you spoke of transcending the subjective and objective to reach totality. I disagree, I would rather seek to merge these two aspects to the point where the "I" is no longer conscious. Thus, one attains a true state of objectivity like a child, eyes wide open, curious, receptive, no thought of I will look for this, ignore this. To be, in short, non-prejudiced toward experience. This objectivity is a state of love such as referred to in my article on "Science and Art as Creation," and as expressed by Russell, Einstein, and Einstein about Planck. It is this experience of love that I believe can lead the way to a sense of true love between humans. In the case of a parent this requires a period of non-parent consciousness, that is, absence of the conscious "I."

Because we do not see alike on the above problem probably explains your rejection of the attitude you assume I have toward Picasso and the rest. Analysis of these forms of art has been based on the effort to remove the conscious "I," but love is circumscribed to the analysis, not to these artists. To accomplish this I have even projected myself into painters like Picasso and others, by means of having their art before me. I also do this with past artists whom I admire, where love this time for the artists has, too often, blinded me to their defects.

To what extent does the wide difference between your field and mine influence each in his conclusions about the above? You physicists in your personal riffs have only each other to deal with, and the words each of you utters about nature. Nature, your aspect of nature, however, remains imperturbably out-there, where each of you knows none of you can change it. I mean, none of you can start ignoring atoms and peopling nature with xerdons, whatever that is. In art our nature is not held so steady out of the reach of monkey business, we don't have instruments to check crackpot antics with nature. This situation makes greater demands upon the artist in pursuit of objectivity of nature, than in the case of scientists, believe it or not.

After repeated readings of your revised remarks on identity, nullity, group theory, etc., I fail to reach your thinking. It is due to my failure to think in terms of identity and nullity which, therefore, form blockages to my understanding you.

January 18, 1962
(Reply to yours of December 26, 1961)

When you say you have not seen much Structurist art, then I must correct you. I originated that name in 1952 to designate not a school but a historical evolutionary direction of a creative art. I recognize no *living* artist as a Structurist outside of myself, with two exceptions: artists who have yet to come to a place where one can say they are Structurists in the sense I comprehend it. So, unless you refer to my work, you are not referring to SA, but to Synthesist, Constructionist, Constructivist, Neoplastic, and the like. I suggest we speak of "new art," to indicate the general effort.

When you refer to such as Hill's art, as cold and calculated (there is also Ernest, Dill and many others), I most certainly agree. Hill is indeed correct to regard himself as a frustrated mathematician as he informed you, and which situation he could easily remedy by ceasing to be a frustrated artist. When an artist makes an arbitrary entrance into the new art, meaning a too self-centered interest, the act of creation churns in the arbitrary waters of the amorphous. Such an artist often chooses to escape to mathematics to give a *raison d'être* to his act of creation. You are right then, to note that the new art attracts not necessarily the math-minded, but certainly the use of math as a crutch. Some of these artists have expressed the view that I have no right to print my criticisms of their use of math, especially since I have frankly admitted to some of them that I am an ignoramus about math. But this should make it all the more easy for them to dispose of me, and I wait patiently.

Why restrict the new to the straight plane, why not the rich curvature which nature expresses? You have not read much of my writing. Here I can only remark on one aspect of this. The assumption is taken that, like mimetic art, purely creative art will sustain its living quality only by an evolutionary development. To do so, one must secure a genesis such as will sustain the pursuit of an evolution. This seems to require reducing the problem of form to its simplest possible terms, the idea being that the simpler, the more potentiality for coherency in a beginning. This is gone about by reducing the sphere, representing nature as basically curvature, to its simplest geometric terms – the cube. (This form later evolves into the spatial plane.) Since the Structurist seeks the example of nature for the structural possibilities of his development, the cube puts him in a position to begin on the simplest structural terms possible. Once again this makes for a coherent beginning, this time as regards the use of nature for the development of a creative art.

Does the artist need to restrict himself to the "antiseptic" plastics, metals, glass, etc.? I must speak for myself, since I reject all these materials, as such. If you were to see my work, you would eventually have to ask me, "What materials are these made of?" For all you would see are colored spatial planes! I look to my predecessors the painters, not to the manufacturers of industrial materials. Antiseptic use of materials, antiseptic *use* of math, do not they go together?

In some future development the new art will have reached a more direct relation to nature as curvature.

Your theory of discreteness and discontinuity in science–nature and art (see my "Nature and Art" article in *S.* [*Structure* – Ed.]). The parallel between the nature discoveries of science and the new efforts of art seem all too evident. Much of this is a deception, as it now stands, all the more easily accepted because of the confusion of new artists before the problem of art, a confusion now of long standing. If we wish to restrict the discussion to discrete structure, I could then say the new artist *continues* on the macro-level where the "jumps" appear continuous. The great change in art has not been one of, can I say, form beyond form (discrete), but only from a mimetic to a different kind of form – from mimetic (nature) to creative (man) forms. I do not wish to deny parallels between the developments in science and art, rather I expect they will be really interesting ones, not merely reflective but creative. The latter will throw unique light on art and science.

All through the development of a new direction for art, science has periodically been sought as the Salvation Army for the artist. For instances, right with the beginning, Impressionism, artists looked to the scientific view of color. Monet, the greatest Impressionist, never did, and Pissarro had to eventually abandon the science color. The problem of color in art was unique, obviously it was not that of science. Cubists and de Stijl artists both used science theory to rationalize their art, along with math, as is still being done today. Cézanne, the first and greatest Cubist, solved that problem *with art* before nature. (See my "Art and Philosophy" article coming in *S.*)

Are my abstractions in art "natural" or "personal?" There are two levels or orders to the answer. In one we have the art which determines its own course of the *particular forms* that take place in its development. In the other we have nature which offers the field of structural possibilities, the only possibilities, for the realization of art and its evolution. Art is thus the creation of man according to the infinitely rich creative potentialities made possible by man. (See *TS* [*The Structurist*] no. 2, which I will send you.)

In questioning the situation in which my efforts at analyzing art may exist you mean, to put it more bluntly, that it is easy to see what we want to see. OK. So, do I ever throw up everything and begin all over, in order to avoid the danger of conditioning? This is unavoidable in our changing-over times. From 1926 to 1937 I went through a constant series of such experiences, as a painter. From 1937, when I became what I now call Structurist, to 1949, the period of my *Evolution* book, I went through a number of serious changes which revolved around the two schools of de Stijl and Russian Constructivism, and which period displaced these schools with a new view I formulated of Cézanne. Thus, I was almost in my mid-40s when I found what I was seeking for. The last 12 years have not contained further upheavals, nor do I expect any.

Am I choosing forms "rich enough" to accomplish my desired end? You will grant me that I could easily make my forms "rich enough," all you want, as do the followers of the Constructivist Russians. But I have answered this in preceding comments. Here I could ask this. Why didn't Bach use such a "rich" composition of sounds as Beethoven? To create is to learn to create. To learn to create is to establish a sustained process of creative evolution for art. This is sensible, but better, experience confirms it. Why didn't nature start out with its rich forms? Why did it perhaps begin with an amorphous gas in which the particles were created, from which the galaxies were created, etc., etc.? I am not one of the artist–heroes of our century who thinks to do better than nature. To create is to grow, not to pick something ready-made off a store-shelf. The artist who supposes he is a nature unto himself, wanders in the delusions and hallucinations of his own puny self.

What you say about the consequences of my published writings is more correct than you can possibly know. I became well aware of what happens to published writings, especially theory, within a few months of my first book. I paid for it for not being prepared for it. What you say is "liable" to happen, my vision being confused and lost at the hands of readers, has indeed happened. The now so-called British Constructivist school is the prime example of it, which school you may not know began as the result of my published *Evolution*. I am opposed to forming any school for the new art, and a nationalist one is beyond forbearance. No, none of these experiences made me withdraw to Red Wing, but it does make me content to stay here or some other such place out in the clean vision of nature.

You are very correct to state that the central question is not mimeticism or non-mimeticism, as oppositions. Along your similar lines of reason, one cannot achieve democracy on the basis of anti-Communism as certain elements in the US are proving all too clearly. In the same way, one cannot achieve an art of creation on an anti-mimetic basis. I know because that was the kind of first attempt I made as a painter. For this reason I have often stressed, and I do this particularly in the "Art and Philosophy" article mentioned above, that mimeticism legitimately continues, but in another medium. It is also the reason why I gave so much importance to photography in my *Evolution*, that book being the first art history to give a legitimate place to the camera in art history. As for mimeticism in painting or sculpture, that practice has achieved chaos in trying to go forward, and there is no return to the past. This is only to emphasize that mimetic art must go forward like anything else.

Can we liberate ourselves 100 per cent from mimesis? That is not at all difficult to do, anyone can do it. The pertinent question is, "Have we liberated ourselves to something or nothing?" But liberation implies anti-mimeticism again. So is there the possibility of a purely creative art? Visual creation is not entirely a new thing. Architecture and all the applied arts have exercised this effort from the beginning of human culture. But, in the applied arts creation could never be more than a limited expression of creation. So, the question really is, "Can there be a pure form of art creation?" But here it is of no use to use words if you cannot see it, and you cannot see it without a new vision. It being obvious that a mimetic vision will not suffice. The same arguments and problems came up with the divorce of music from literary functions, some almost 300 years ago. As we cannot listen for the story in music where only the creation of sounds stands alone, so we cannot look for the story and see the art where there is no longer a story.

Your exposition of independent orders is logical and impressive. Yet I am not convinced. The greatest power of man also contains his greatest danger, namely, that it is possible for him to set up situations, such as experiments, which do not and cannot occur in unmolested nature. Nature does not toss coins. I go along with Einstein, that nature is not running a gambling casino, no dice-throwing. Anyway, does man with his coin-throwing set up a circumstance of structural happenings that are not sufficiently understood, and therefore give a false report about the structure of nature? In any case, is it possible to show beyond the point where justifiable dispute is possible, where fact is offered rather than faith asked, as I think seems to be done in the disorder–order notions *about* nature? If the latter could be shown the arguments would end, and there would be no necessity to appeal to the coin-throwing, about which we would then perhaps know more too.

I find it difficult to accept notions of randomness, disorder, dislocation, distortion, and the like. That seems to me, to put impressive lingo at places of our ignorance. I believe it to be of first significance that in science and art (both the mimetic manipulators and those who deny nature), that both speak of nature in terms of its having certain defects, and this precisely in a period where members of both fields speak of a "crisis" between man and nature. Terms like random, etc., appear to me like notices posted over certain areas of nature reading, "Here anything can happen, feel free to speculate whatever you think." These terms are mean-nothings which everyone is free to make into mean-somethings, an arbitrariness, a surreal indeterminism. It reminds me of Hans Arp. He would make an ink drawing and then tear it to pieces. He then drops the pieces on a sheet of paper. And, wherever they happen to land he pastes them down. Thus, a finished drawing. Perhaps if I were not so very ignorant about science, I would know better than to say what I have, or else I would be able to better say what I stumble around to say.

Your illustration of the molecular theory of gases interests me particularly, but I could not *visualize*, therefore I could not comprehend, the crux of your exposition. Could you make me a drawing showing molecules "moving regularly and nearly in the same direction," so I could see how they would collide and make a turbulence of irregularity on the macro-level, and a drawing of the random movement on the sub-macro? I am enclosing something I wrote almost eight years ago, that is related to our discussion. I sent it to the mathematician E. T. Bell, with whom I have corresponded, and also to W. H. George at the Chelsea Polytechnic. I would like your frank opinion of it.

Your remarks on symmetry make me very curious but unable to cope with them. Perhaps you will say more about it, something that my novice mind can grasp. This reminds me, in the early 1950s I had some correspondence with Kathleen Londsdale on symmetry, randomness, ideal models, etc.

Were you able to see my actual art and arrive at your own actualized experience of it, you would then realize how very correct the remarks in your last paragraph are. I mean where you say that, in regard to the new art you have experienced, the big problem is the question of contingency and necessitation from which arises that "inner tension" by which the art comes alive. Finally, you say, art cannot be illustrative of math relations. Discerning this, which most new artists are very ignorant of, should facilitate your discerning the necessity for some proper beginning for a creative art such as can sustain an evolution. If the living quality of the art is that which is poised in the tension established between the forces of necessitation and contingency, the necessity for the growth of all these factors can only be realized by some coherent direction of evolution. This can only be accomplished if the creative activity in the art is consciously related to the study of nature's creative aspect, for there exists the supreme manifestation of living structure. It is astounding, I have often referred to the living quality which is necessary in the new art. Yet, not one artist has ever mentioned it to me.

Is it any wonder that when the artist is unconscious of what is indeed his inescapable relation to nature, that he unconsciously experiences a lack of the living quality in his art, that he then tries to remedy this situation by turning to such a time-honored discipline as mathematics? To the degree an artist can be a mathematician in his art, to that degree he deludes himself *as artist.* Because he will mistake his *feeling* of mathematical accomplishment with the feeling of a living quality for art.

Your remarks on mathematics as creation are the first profound things, to me as an artist, that I have heard about mathematics. Here we have the opportunity to establish a sensible relation between art, music and math. But, if there is math in art and music, is there art and music in math? Perhaps it is no mere accident that some mathematicians and physicists have felt they are artists. What is an accident and serious is that some scientists, well-known ones, have encouraged artists into calculation, "the result of technique," as you put it. What you say about the technician and calculator applies to such artists. There is a void between themselves and art, the latter being like a foreign body outside himself. There is lacking a living relation between the artist and his art. He does not, he cannot, feel himself to "be it," as you say. He is not one with art. Thus art serves the artist in much the way the calculator's work serves him, to gratify desires for fame, wealth, etc. We should note, however, that to the opposite of lifeless art there is another that is indeed alive. It literally seethes with the living quality, but it is the seething life one can experience before a psychotic person – the art of pathology, the mimetic destructions of nature.

Yes, I am married and we have an 18-year-old daughter. We send you and your wife our warmest regards.

Charles Biederman

3, Berkshire Road
Bristol 7
Glos., England
December 29, 1961

Dear Mr Biederman:

This is a very brief letter to summarize and add to a few points that were made in my previous two letters (Dec. 22 and Dec. 26).

You are quite right in saying that we have now reached the problem of defining our terms, of getting to the basis of a real understanding, by going into the question of how we use words.

As I see it, the basic problem is of what is meant by "same-ness" or "identity". Following Korzybski, you say that no two things are ever really the same. And here, I agree with you, up to a point at least. Then you go on to bring in the idea of multi-ordinal terms (for example, with redness, symbolized by R, you consider R_1, R_2, R_3 . . . etc.). But here I raise the question, "Why use the same symbol R?" Why not replace R_2 by B for blueness, S, for sweetness, etc.? In spite of the fact that R_1 is not R_2, they must be the same in some respect, or else there would be no justification for using the same word, and language would turn into utter chaos.

I proposed to solve the problem (at least in part) by bringing in the idea of a field, and the concept of the nullity of a field. Two things (that are different, of course) may be the same inside a certain field, differing by the nullity of that field.

This raises the problem, of course, as to whether or not the projection of the world into certain fields is not arbitrary. But you yourself have in effect provided the answer to this question. In general, such a projection is, of course, arbitrary. And the arbitrariness shows up in fragmentation, disintegration and confusion, which result when you project into the wrong field. But then you considered the idea of a "natural" abstraction, which does not produce fragmentation, disintegration and confusion. On the contrary, "natural" abstraction brings out the essential relationships in a given totality, by stripping away all that is not relevant to the totality, all that is accidental, and contingent in relation to this totality. But the important point is that it *leaves* the essential without breaking it, distorting it, or destroying it.

In order that such process of abstraction shall be possible, it is necessary that the process of the world has a certain general kind of structural relationship, which I can call by the name "abstractability" or "projectability". That is to say, there must exist in the process of the world as a whole certain sub-groups or fields, which move into themselves, in a certain kind of correspondence with the overall process. In other words, the relations in the sub-groups reflect certain relationships in the whole world process. But what they must reflect is what is essential to the totality of the world process. Otherwise, the abstraction will be a trivial one. This means that they reflect the very process in which the elements of the world totality are generated, in their actual relationships. In this regard, even the abstracted group or field itself is an element or aspect of the totality that it is abstracting. And here is where your stress on Structurist art as an integral part of nature structure comes into relevance. In other words, a piece of art need not be just a mimesis or reflection of nature, it is a part of nature, an aspect that could not have existed without man's creative activity. The weakness of mimetic art is that it is only the content of what is reflected that is relevant, and not the process of reflection itself. A Structurist art should then be an abstraction which is plainly part of nature in its own right, and yet also an abstraction or a projection of what is essential in nature's totality. It should also have in it the whole story of its relationship to nature; i.e., that it is actually an abstraction of this particular kind. Am I right in ascribing these views to you?

This brings me to the idea of the concrete. Now, is the concrete just the opposite of the abstract? Evidently not. For if it were, it would be that which is not abstract, and therefore only an abstraction from the abstract. This would make it doubly abstract. Rather, the concrete is the existent actual totality from which all abstractions come. As such, it contains the abstract as a part or aspect of itself. In other words, the concrete is (among other things) also the abstract. But this abstraction is in reality a concrete abstraction in the same sense as Structurist art. For the piece of Structurist art is an aspect of concrete reality in its own right, and is not just a mimesis. And yet, it is an aspect in which there is a field, containing essential relationships in the concrete totality. As I said in my earlier letter, it is an abstraction which is always plainly saying that it is just what it is, a part of reality that is abstracting something essential from the totality.

If you look in a mirror, you see a reflected world. The content of the reflection is the world, but then there is the process of reflection itself. Usually, one is not aware of this process, which is a complex movement by which the mirror reflects the incident light in a certain order and relationship to the sources of the light. In other words, you hardly pay any attention to the mirror itself, but you are interested mainly in the reflected image. But now the time has come, in science as well as in art, to pay more attention to the mirror, and to the process by which it reflects the world. We must notice that the mirror is itself an integral part of the world; and we must find mirrors which are able to reflect essential features of the world process, including the fact that they are mirrors related to totality in a certain way.

The problem is much more general than that of science or art. As I indicated in my earlier letters, it has to do with understanding and truth. For understanding is perception of totality. And I think that what is perceived is just the essential field, which is appropriate for abstraction this particular moment. Thus understanding goes from moment to moment. Truth or falsity are also in this process. Truth is, I think, non-contradiction in the total field of experience, including perception, thought, feeling and action. Falsity is a process of contradiction of one part against another, either one thought against another, thought against feeling, feeling against feeling, thought-feeling against action, or action against action; or any of these against perception. There is therefore no final truth or abstract truth, but rather, a concrete truth, which develops from moment to moment along with understanding, as the total situation changes. Truth is not just something wholly outside ourselves, which we approach step by step; nor is it wholly internal and subjective. Rather, it is something that comes into being within us, by which we are related to totality. This totality is neither purely internal nor purely external, but contains both. As you say, we are in it. And through truth it is in us. In a moment of creative understanding, we are truth, and when we are confused, we are false, and even more, we are falsity itself, insofar as we accept and identify ourselves with a process that is false.

We then come to the problem of contingency. This arises because there are many fields of abstraction, different subjects, different moments, etc. These fields have, as I explained in my previous letter, a certain independence, such that one does not completely determine the other. Thus, given what is in one field, developments in another field are contingent. But because fields are not in general absolutely unrelated (except in the case of random relationships), the lack of determination of field B by field A will reflect back as a lack of determination of A by itself. For in some ways, A depends on B, and B is in some ways independent of A. We therefore must have contingency in each field.

Notice that I am not saying that any given event is pure chance. I am saying first of all, that the existence of fields of abstraction is characteristic of nature's structure and order. The partial and relative dependence and independence of these fields is also characteristic of nature. Therefore, in each field, there must be contingency and chance. What is chance in one field may be necessary in a broader field and vice versa. But all thought functions in fields. Therefore, we will never get rid of chance and contingency in our thought. But this is not purely the result of our own way of thinking. For this way of thinking is based on the existence of fields of abstraction in nature (what you call "natural" abstractions). There is genuine contingency, insofar as functioning in fields is valid.

Moreover, I further claim that necessity is neither more nor less real than contingency. For necessity also operates in some field. Thus, you can say that in a certain field, a given development is necessary. But in general, you find that each field is related to a whole set of broader fields, having a certain independence and dependence. In physics, this is very clear. Thus, if you have a gas, consisting of many molecules in random motions (in relation to the large-scale level), then it is quite necessary that if you open a valve, the gas will flow out (if it is under pressure, of course). If you analyse this process in a broader field including the micro-level as well as the macro-level, you will see that this necessary regular behaviour is the result of what are, in this broader field, a host of accidents (e.g., collisions, movements of particular molecules to and fro, etc.). Similarly, the regular functioning of a city with regard to traffic, food supply, etc., depends on the co-operation of myriads of plans, aims and desires of individual people, none of which is very directly related to the overall traffic pattern.

The fact that necessity dissolves into accident and chance in a broader field is crucial to the possibility of freedom. For every chain of necessity, however iron-bound it may seem, is seen in a suitably broad field, to be full of holes, dependent on accidents, etc. Therefore, if we understand necessity deeply enough, we can always escape it or modify it. This is an important idea, a sort of extension of what you are driving at in Structurist art. For nature as understood by man is different from nature not thus understood. Causality operates in an iron-bound way only to those who do not understand it. In this regard, human nature being part of the totality must understand itself as well as external nature. Thus, again the relationship is reminiscent of what you say about Structurist art. Human nature is not just a reflection or a mimesis of external nature, it has a concrete existence in its own right. But it must also abstract the totality, including itself, if it is not to be petty, trivial, fragmented, disintegrated and confused.

Of course, you might say that each event is determined necessarily in a complete totality. But this would be trivial. For to give the complete totality, you would have to give each event, so that the determination of that event wouldn't add very much to it. Rather, necessity and determination have real content, only insofar as they operate within fields of abstraction. And it is an essential aspect of nature structure that this type of necessity–contingency actually does operate in every such field of abstraction, even in those that are "natural".

So it seems to me that the process of contingentation–necessitation that we see in art corresponds, in its own way, to relationships that exist in nature structure. We do not introduce contingency into nature. If we consider nature without us, this is an abstract field, which therefore necessarily has contingency in it. For this field depends on what is left out, which is in this case, just ourselves. But such an abstractability into "natural" fields, each having contingency, characterizes nature in every aspect. Contingency is an essential part of the true relationships in nature structure (being in this regard neither more nor less essential than necessity).

Now, the meaning of contradiction is this: each partial field will have contradiction, if it is taken to stand by itself. Reality is the implicit. It is that which produces these abstractions, these partial fields, these projections. It is a totality that cannot be put into separate words or fields. But the meaning in its totality can come into being in each person, just as the meaning of a musical composition does, even though the separate notes do not mean much. Here, Korzybski's idea of multi-ordinal terms is relevant. For each word has many possible meanings (even an infinity of them) and the actual contribution that it makes is determined only in the light of the total idea. Nevertheless, each word is in some way limited in its meaning, or else it could be freely exchanged with any other word. In some essential respect, each word is at least relatively fixed to a certain range of meanings. By "meaning", I wish to suggest a certain function in the totality of the process of thought and communication, at least during a certain moment, or during a certain period of time.

Of course, the world is not words or thought. And as you say, we have to understand the way in which it can be abstracted, by a process of structural correspondence, into words and thoughts, which latter are part of reality. I think that we have to understand the true meaning of "sameness" or "identity" if we want to do this. Identity in the simple sense of absolute sameness doesn't exist. Yet, it has a real meaning, and plays an essential part in the function by which reality can be abstracted into words and thought. If one doesn't understand the real meaning of identity, one will come to contradiction, and eventually this will lead to conflict in perception, thought, feeling and action.

The letter turned out to be longer than I thought, but as you say, this is a key problem which we must get clear. I am waiting with great interest to hear what you have to say with regard to these three letters.

Very sincerely yours,
David Bohm

3, Berkshire Road
Bristol 7
Glos., England
December 30, 1961

Dear Mr Biederman:

I would like to supplement yesterday's letter, as I have been thinking a bit since I sent it off and have had a few more ideas. (The vacation is now coming to an end, so I probably won't write much more for a while.)

What I am thinking about is the problem of "terms" in language again, with the question of what is the meaning of identity.

You suggest in your letter of May 29, 1961, that we should begin with undefined terms and not with definite terms. With this suggestion, I am in general agreement. Yet there is the problem of why you can use any terms at all, defined or undefined. Consider, for example, your suggestion of such a treatment for words like reality, democracy, science, art, socialism, atom, nature, quantum, etc. Thus, there are many cases in which we use the word democracy, which we represent by D_1, D_2, D_3, D_4, etc. Another word is fascism (F_1, F_2, F_3, F_4, etc.). We cannot correctly put, for example, F_3 in place of D_3. No matter how the situation changes, there is some essential respect in which democracy remains itself, even if we can't exhaustively define what it is. I propose to treat this problem by saying that with regard to the essential meaning of the term "democracy", the difference between D_1, D_2, D_3, D_4, is not significant, at least compared with the differences between democracy and fascism, or for that matter, between democracy and atomism, democracy and quantum, etc.

As I said in earlier letters, I think that the definition of such terms is only by implication. The explicit statements that we make are only abstractions or projections, and the real meaning is that implicit process which has these abstractions or projections. The word is always changing its meaning as we use it. We both agree on this. Yet, I think that in some sense, we understand or appreciate a concept in its totality, before we appreciate its detailed meaning. By totality, I also mean the essential process, which cannot be explicitly defined, but which can be perceived from one instance to another, just as one can perceive that the differences between D_1, D_2, D_3, are inessential to the basic (implicit) meaning of democracy.

So I think that the notion of identity as a relationship of difference by the nullity of a certain field, or by what is inessential in a certain "natural" kind of abstraction, makes an important contribution here to clarifying what Korzybski was trying to express.

Your suggestion of "similarity", for this purpose is, I think, not adequate (as I indicated in earlier letters). For the idea is too vague. Thus, everything is similar to everything else, as well as different. Redness is similar to blueness in being a colour, it is different in being another colour. Redness is similar to itself, yet, as we agree, each instance of redness is different from another. Democracy is similar to fascism in being a form of government and it is different in other obvious respects. Democracy is similar to atomism in that both have a philosophical content. So as you see, any two things are both similar and different. Yet certain things have the right to be labelled in a given sequence D_1, D_2, D_3, etc. while others are put into another sequence F_1, F_2, F_3, and C_1, C_2, C_3 (Communism 1 for Russia, Communism 2 for China, Communism 3 for Albania, etc.).

Now the main point here is that with regard to a certain "natural" field of abstraction, certain differences are essential, while others are inessential. In other words, similarity by itself (as well as difference) is not a very significant idea, unless one includes also the respect in which two things are similar (or different). And this respect can be trivial or essential. So one must include the idea of essential similarity and trivial similarities. By the time you go thus far, you will have to bring in effectively the idea of two things being the same in certain respects, then differences being the nullity of a certain field of abstraction. (Trivial differences being differences by nullity in such a field.)

This raises an important question. Is a respect similar to itself or different? For example, if redness is similar to blueness in the respect of being a colour, is this respect identical with itself? The answer is, of course, "Yes and no." For each instance of colour is different. Yet these differences are in the nullity of a certain field, so that the respect of colour is, in this sense, similar to itself. This shows up clearly the problem that is in all language. I think the solution lies, as you suggest, in recognizing that all terms are undefined, along with what I suggest, that their meaning is implicit. However, I feel that thus far, you haven't paid enough attention to understanding the meaning of what is usually called "identity". I admit it isn't really identity, but I think there is an important problem here, not adequately dealt with by Korzybski's system of labels (D_1, D_2, . . .).

Perhaps an important source of misunderstanding between us is in the question of whether as you say in your letter, under certain circumstances, differences appear not to affect a particular observation of structure. I think it is not just a question of appearances but of actuality. For in a certain "natural" field of abstraction, their differences actually do not have any effect. In other words, I think that besides "thresholds of perception", there are natural, real, objective relationships between different fields, such that certain happenings in one field have effects which are below the threshold of another field, or which for some other reason do not function in the other field.

Coming back to the implicit meaning of all fundamental terms, I would like to add that a certain part of the whole set of relationships of terms can be made explicit. For example, even though a term like democracy cannot be fully defined explicitly, it appears to represent some kind of totality, in the sense that there is a "natural" field of abstraction, in which it can be perceived that the differences between D_1, D_2, D_3, etc. are inessential to the basic relationships which make democracy what it is. In this connection, we must ask ourselves what leads us to bring in a term such as democracy in the first place. It seems to me that we feel (at least intuitively) that there is a certain total process here which really does define itself and distinguish itself, in the sense that I have just described. For example, we would not put the term "fly-speck" in the same category as democracy. Have you ever asked yourself "Why not?"

With regard to the definition of terms, I don't remember whether I said in earlier letters that whereas all definition is implicit, we can nevertheless explain or explicate (i.e., make explicit) certain limited aspects of the implicit meaning of any term. One does this in order to establish a few salient points, with the purpose of setting up in the mind of another person essentially the same process as in one's own (i.e., the differences must be in the nullity of a "natural" field, defined by the problem itself).

There is evidently much that is not yet clear in this question of language. But I agree with you that we must try to clear it up. I have tried to answer directly your responses to my views, as you requested. The answer has been lengthy, partly because the problem is complex, and partly because my expression is unwieldy, because I am still developing my ideas on the subject. Now I await your further responses to my responses with very great interest.

Best regards,
David Bohm

P.S. One more point about the relationships of mathematics, music and art. You ask why mathematics has not been helpful in composing music. My answer (in previous letters) was that you cannot calculate the creative process (you might equally well ask why mathematics cannot be used in the very process of inventing new kinds of mathematics). You also ask why those who do not compose music are interested in applying mathematics to the art. I really don't know, but perhaps it is an effort to find a substitute for the ability to compose music. In other words, they enjoy music, appreciate it, but are for one reason or another not capable of creative activity in this field. Mathematics seems to provide some control over the field, without the need for this kind of creativity. Perhaps similarly with regard to art. If this is the case, then these people are fooling themselves. In other words, while it would of course be useful and interesting from many points of view to study mathematical relationships in art, music, and even in the foundations of mathematics itself (which are called "meta-mathematics"), such a study is no substitute for actual creative work in these respective fields.

Your discussion of your own experience in the creative act is very interesting, and I hope to say quite a bit more about this in another letter. "Maintaining verbal silence", "putting aside all I know", "experience of totality", not as knowing it, but as being it. All that suggests that we have to get out of the level of purely verbal abstractions. And certainly, we must bring in pre-verbal experience, as you suggest. Even mathematicians find themselves doing this, as one can see from mathematical terminology, which is often quite picturesque. Thus, there are singular functions which "flow up", "act wildly", "pathologically", etc.

But perhaps there is even more to it than this. For there is the experience of understanding and truth, which evidently can never be fully referred to any other kind of experience at all. (I think that this is the basic characteristic that is needed for intelligence.) Such an experience contains thought, feeling, pre-verbal experience, impulse to action, and all the rest. But it is more than any of these or all of them put together. For it is the integrated totality of inner and outer that comes into being from moment to moment. This totality is a process that generates all these other aspects as sides in their proper relationships, but in its aspect of being a concrete unique totality in a given moment of existence, nothing else can explain it or take its place. Rather, all explanations start from the presupposition (implicit) of such an understanding. If you once understand something in its truth as the implicit, then you can later explain what you understand (make it explicit). But if you don't understand, you can't explain, and no amount of explanation will substitute for understanding. Indeed, the purpose of explanation is generally to communicate your understanding to someone else, but in so doing, you will develop it, see it in a new light, criticize it, etc., and eventually, you will see it in the light of someone else's understanding.

Now, a few words on contradiction again. I think we agree that contradiction will arise wherever we try to use terms in a well-defined way. For example, consider the words "similar" and "different". Both of these are names for relationships. They are similar relationships in many respects. Both relationships apply to everything. Everything is similar to everything including itself, and it also differs from everything including itself. Thus, they are similar in the respect of being a certain kind of relationship. But they are different in precisely the same respect. Indeed, even the use of different words indicates that we mean them to be different. And this difference is not trivial, but essential (for the difference between similarity and difference is evidently very fundamental to our understanding of anything whatsoever). Nevertheless, you will find that in the same statement, in the same breath, you must assert their similarity and difference in the same respect. This is a contradiction. And a "similar" contradiction arises whenever we try to define our terms.

We agree that reality is not contradictory, because it is not any kind of words or thoughts at all. But you ask whether it is necessary for the contradiction to arise in our thought and language, in our effort to abstract reality verbally and conceptually. I think we both agree that there is a way out, if we recognize that separate concepts and words must not be too sharply defined. The meaning is in the totality of what is said, so that when we try to localize it in single statements, we distort it, and come to contradictions within each statement and between statements. The notion of many meanings to a term (T_1, T_2, T_3, etc.) helps in this, but then there is the deeper question of how one kind of term, T, splits itself off, in a "natural" way, from another kind of term, S. I have made some suggestions about this, and would appreciate your ideas on this point. But I would like to think of meaning as containing a total structure, including various kinds of abstractions, which in some way reflect the total structure. Certain kinds of abstractions yield a "true" reflection and others a "false" or distorted reflection. This point needs more careful study too.

Red Wing
Route 2
Minnesota, USA
February 24, 1962

Dear Bohm:

Your last two letters seem to have moved me more toward your point regarding the implicit character of what you call identity, but without being able to label this with the term identity. Granted that the term "democracy," for instance, always calls up a quality of implicit meaning which is automatically, silently assumed by many, not all, and that from this implicit base each individual proceeds to generate abstractions about democracy, identity exists only when the individual deliberately or subconsciously identifies. This identification can only exist as a verbalism, what AK [Alfred Korzybski] called word-noise, having no basis in actuality.

When identification is avoided, the implicity of the term democracy is constantly changing for the individual. If this were not so, the abstractions generated from the use of the term would remain generally similar to previous utterances of the particular individual. Identification tends to congeal the implicit aspect into the confines of some absolute. This brings us to what I consider the fundamental nature of the problem you bring up, namely, that the implicity of the term contains our *basic assumptions* about our notion of actual democracy. It is from this *silent or conscious* assumption from which we prescribe the *area of our abstractions* about democracy. For instance, my basic assumption would exclude from the legitimate area of abstractions, that well-known label used in China, "the people's democracy."

Anyway, the "implicit process," it seems to me, is more precisely labeled as the assumptive process which, if we respond to the flow of new facts in experience, is being modified and so our abstractions are being modified too. Because of the crucial necessity of this general process of experience and language, all connotations of identity are harmful and frustrate the multi-ordinal attitude. Subscripts, incidentally, would include time–place, D_{1962}, D_{400BC}, etc., in order to differentiate between the changing notions of a particular individual, or the changing notions evidenced in historical persons, and the like. Agreeing with you that the implicity of key terms can be explicated, I would say it was the basic assumption they contain that gives substance to their explication.

True, the moment I say to you, "I want to speak about democracy," at once you are aware of a totality out of which I am presumably going to speak, even though you are not aware of my basic assumption. What takes place here? First of all, you assume the term will be used as it is currently expressed in the dictionary. For this reason we would not use the term "fly-speck" in the same sense as democracy. There is general agreement to at least use as our starting-point language labels agreed upon by general usage. Hence, the dictionary serves as a verbal convenience. If now I follow my above opening sentence with expressions of approval for the "people's democracy" of China, we have left the realm of the dictionary, and are now on our own confronted with our respective basic assumptions about democracy.

If our language process is to seek a structure corresponding to the known structure of nature, which structure does not display identity, not even in its simplest bilateral symmetries, it follows that identity does not exist except in mere words. The necessity of the term similarity then becomes evident. It implies the rejection of sameness (identity), it implies that important differences are to be expected. This is especially important in a field where seemingly similar basic assumptions may occur. One also avoids contradictions, that is, referring to "the meaning of what is usually called 'identity'," and then having to add that "it isn't really identity." (How do you resolve this contradiction, or is it that you have and I have missed it?) It seems to me meanwhile, that however thin you slice the notion of identity with quotes, it has only the verbal function of frustrating the actualities of process orientation. Identity is the sort of verbalism that compels a racist to concentrate attention on the poor or criminal Negro, pretending Negroes of accomplishment do not exist. The former readily feeds his static, elementalistic basic assumption of the identification of what Negro *is*, the latter frustrates the identification activity.

I agree with your criticisms of my term "appear," regarding situations where differences appear not to effect particular structural observations. While agreeing to all you say, I would still prefer the term "appear" for two reasons. One appears in the latter part of the paragraph where the above appears in an earlier letter, namely, that one remains conscious of the areas of nature that nevertheless surround the particular threshold under consideration. Example: imagine what a tremendously different cultural milieu we would each have to live in and to live from, if all artists kept science (taught to them by competent scientists), in the background of all their abstractions within the threshold of art, and if scientists did likewise with art. It would be the biggest revolution ever in the whole history of man. Second, much like the first, by responding in terms of "appears," rather than in terms of "do not have any effect," is to adopt an *attitude* of tentativeness in place of finality expressions, thus remaining psychologically open to the revisions that will certainly appear, sooner or later.

Your remarks on mathematics, art and music are very interesting. They suggest that you are on the track of discovering means whereby the extant confusion over these issues can be clarified. Would I be correctly following your line of thought, if I said that mathematics cannot be the means to creation in mathematics, art or music, because it is only a particular subject for creation? Your explanation of why mathematicians may apply mathematics to music – substituting the control of mathematics for want of the control of creation (you do not put it so, but that seems to be implied) – as similar to the effort of those who apply mathematics to art, corresponds to my own impressions. (By now I have seen this happen to a number of artists in the new direction, including those who display no evidence of having a true regard for mathematics. It serves merely as a lifebelt.) Is not this why a work of art mathematically determined exudes a sterility? Important too, in art the eye does not see mathematically, not even in the more simple, more gross metrics of arithmetic. It is quite possible that this act of vision can be accounted for mathematically, especially given the precise character of the new art direction. But when this is discovered, will it help any more than all the mathematics that has been applied to explicate the music of Beethoven? Still, who would have thought, until men like Alberti and Leonardo did it, that the human image could be accounted for with the tools of geometry and mathematics? In this connection, however, artists eventually ceased to literally apply the laws of perspective, regarding them only as a background while depending on their vision. Are you familiar with William H. George, *The Scientist in Action*? He sees the work of mathematics continued by the method of "patterning," as in crystallography, which he regards as closer to the method of the artist.

I find your discussion apropos the creative act very good. Certainly without the integration into totality of both inner and outer, creation can only be confused and fragmentary, if not pathological. Would you agree that it is not that anyone escapes some totality orientation, but that given the tricks of language coupled with the ignorance of language as mere verbalism, it becomes possible to conjure totality largely based on verbalism which has a fragmentary, narrowing effect upon experience and action?

About the contradictory relations you establish between the use of the terms "similar" and "different." True, both are names for relations, both imply similar relations in many respects. But I would also add, different relations are also implied. Everything is similar to everything else, in the context of process structure, but not (identity) including itself (an object, particle). The latter is always becoming different to what it was (itself) while remaining similar to everything else. Hence, the relations of similarity are structurally limited, while relations of differences are structurally unlimited, in this context. It does not then follow that you must "assert their similarity and difference *in the same respect*," in all cases. The above replies to your Dec. 30; what follows is to your Dec. 29.

Your discussion of fields is helpful to me, I must think more in this way. You lead this discussion into Structurist art as an example, and then ask if what you say expressed my views. Yes, particularly on what you say about S.A. directly, I would not discriminate mimetic art as "just a mimesis," as "only the content of what is reflected." If the Structurist direction establishes a deeper relation of art to nature, it does not achieve a substitute for mimetic art, or cancel out its usefulness. Mimeticism is a unique form of visual experience that is far from exhausted. Thus it has naturally happened that it not only continues, but continues an evolution as that can be done, meaning camera art in all its forms.

The great mimetic artists of the past remain unique in our experience, perhaps in the same way that you receive certain experiences in regarding the achievements of a Newton. But most important, the past artists were not, because they could not be, "just" mimetic. They did not depict "only" or depend only on the content, in the sense that they immersed the human being into a literal embrace with nature. They did not merely reflect nature, their art reflects the meeting of man with nature, and that alters everything, gives it meaning which would be absent if mere reflection was involved. True, as art evolved, the capacity to reflect nature evolved, but this was always the occasion for a new and deeper meeting of man with nature. The fact that the genuine mimetic artist of today uses a camera (machine), means that the art medium has been perfected, as has been the case in the past, in order to continue the evolution that has taken place in the past, again to new experiences. Many mistakenly believe that the camera represents the end of this road – far from it, it is a new beginning. From this view the art of mimesis can be seen as a *particular* part of the "process of reflection itself," a unique part.

Mimetic abstractions, by the way, also remain as a limited case of the Structurist artists' more extensive mode of abstraction. If Structurism creates art entities such as have not and could not exist in nature without man to create them, well this could be as well said of all past art. So where is it unique? In that it does not, in any way, reflect what nature *has created.* Rather, it is a direct reflection of the total creative process of nature. Consequently, *entirely* unique events arise with this form of art in its amalgamation of man with nature, something "in its own right," as you well put it. So the differences between past and present forms of art lie in the different fields of nature reflection, from which each form of art originates and is generated.

Your remarks on the concrete and abstract are very interesting, and I wish you had said more about it. Your notion of the abstract as an aspect of the concrete, therefore a "concrete abstraction," is suggestive as a way to consider pre-verbal or sensory abstractions. Could not one say, that all pre-verbal abstractions fall in a certain order of degrees of concreteness? You will have noticed in my articles, that I speak of S.A. as not representative of anything else, that it is only itself, and like expressions. Your terminology seems to suggest the possibility of speaking about this with more coherence. But the point is, that it is not unique in being derived from concrete abstractions, for these are present also in the making of mimetic art, but rather the point is the one you finally make, that it is "an aspect of concrete reality in its own right." Structurist art is as concrete as anything in nature, as unique as the things of nature. Not only because it possesses all the structural attributes of natural reality, but also because it derives its reason-for-being from the very source of that concrete reality, nature as fundamentally a creative process. Structurist art introduces a new field of the concrete in nature.

Until my discussions with you on the subject, I had not thought about nature as reflected like in a mirror (except in the case of mimetic art), in terms of nature's process itself. It is a valuable way to look at it for, after all, this is just the way I look at it, and the language gets a little closer to the actuality experience. We, however, who are in the transitional period from the old to the new have, I think, the big problem of not carrying over habits of reflection cultivated by the old vision of nature. I never fail to be amused, I sometimes have to laugh outright, when I catch myself rejecting or ignoring some visual aspect of the new reflection of nature or art, due to perceptive habits of the old reflection.

Your discussion on the problem of "understanding" in relation to the true and the false, is certainly to the point and illuminates the problems. In the end it is a question of the amalgamation of the self with totality, as Cézanne might have expressed it.

While agreeing with your observations on contingency and necessitation, their changing roles according to the field or fields considered, still I have difficulty in following that part of your argument establishing contingency in nature. No doubt you have expressed some thoughts there that I missed. If you as a human regard contingency as ever present in all your abstractions from nature, how do you know you are not projecting contingency upon nature, rather than reflecting it?

It is beyond my comprehension that scientists, having attained a very precise relation of their concerns with nature, beyond that in any other fields, will yet be content to place at the very depths of their penetration of nature the very opposite quality of precision, namely a randomness. As I said in earlier letters, how does one recognize randomness in nature, how does one know that it is not a projection upon nature?

When you note that any field will incur contradictions when it attempts to stand by itself, I immediately think about art and science. What holds true of any human field in relation to the totality of nature, is just as true for all human fields in relation to each other.

I look forward to your next letter in which I hope you will continue to have occasion to state your reactions to S.A. Those you have made so far interested me very much.

My best regards,
Charles Biederman

V

1 May 6, 1962

Dear Biederman

I am continuing my answer to your letter of April 14 and April 23.

1 . You are probably right to say that some effort to define what is meant by truth was necessary, and that at least originally, these efforts were not wholly escapes. But, by now, after some centuries, they are functioning mainly as escapes. People get lost in technical details, by which they get results, and thus forget the total activity in which they areengaged.

Why do physics at all? This is a question that is relevant; and yet everyone tries to escape it. At first, a typical physicist may say "It is useful". But is he really interested in its utility? If he were, would he not try to find out whether physics has in fact contributed to human happiness? For example, are transistor radios, television sets, rockets, and hydrogen bombs the real purpose behind what the physicists is doing? Evidently not. If the physicist is not just one of those who wants to make a career in a name for himself, then one finds that he feels it to be worth doing in itself. But he doesn't like to ask himself why he feels it to be worth doing. He prefers to accept his feelings, without being too aware of them. This in itself need not be so bad, but he also feels the need to pick on some justification for his activities. So today he ascribes value to the ability to predict, in itself. He feels that the essence of physics is its ability to predict. Then you can ask him to imagine a situation in which one type of physicist makes apparatus that gives results that are easy to predict and whether this would not give "all round satisfaction guaranteed". Why go to the bother of studying Nature's processes, if one set of men could put out a predictable process, to be observed; studied, and predicted by a second set of physicists? This is of course recognized to be ridiculous; and finally, it is seen that in physics research, man is concerned at heart with the problem of truth and understanding, which is a creative activity. But people are conditioned to find some external, respectable and socially recognized and accepted conventional justification for what they are doing. This creates confusion of thinking, and a resulting idiotic behaviour. Thus, quantum theory is now formulated as if the basic laws of physics were concerned with nothing more than what physicists can observe with their instruments, and as if Nature's process beyond the functioning of these instruments had no relevance whatsoever as far as the physicist is concerned. This all follows logically from the narrow identification of physics with prediction.

To try to define what physics is doing with great precision is not harmful in itself. What is harmful is that these efforts have in fact usually been the outcome of very confused motives; for example, a desire to make the activities of the physicist as recognized and respectable as those of the business man,

BEYOND THE SUBJECT–OBJECT DISTINCTION

3, Berkshire Road
Bristol 7
Glos., England
February 2, 1962

Dear Mr Biederman:

I have received your letter of Jan. 25, and by now, I hope that you have received all my letters. I await your reply when you have the time, with great interest.

In your reply, I hope that you will take into account my remarks on the problem of understanding, as I regard the problem as central to all the issues that you raised. Roughly speaking, I think that the mind functions in two ways. The first of these is associative thinking plus the drawing of conclusions by logical arguments and calculations. In other words, on the basis of many experiences, we learn to associate certain things together habitually, thus forming a concept. This usually happens unconsciously (i.e., without conscious awareness that we are doing it) though it could also be conscious process in special cases. From such associative thinking come the premises from which we draw our logical conclusions, and on the basis of which we make our calculations and plans. In physics, this could be compared with observing many circles, measuring them up, finding approximate empirical formulae such as the one for the ratio of circumference to diameter, calculating on the basis of these formulae, etc., etc.

The other mode of functioning of the mind is understanding. As I said in my earlier letters, this is roughly the perception of totality, in the sense that the parts are no longer regarded as separate things to be put together by associative relationships. Indeed, the totality has a law, and from this law all parts are generated, in a natural way, and in their proper relationships. The example I gave was developing a geometrical theory of the circle. When this is explained, then the hearer suddenly says "I see", meaning "I understand" – and this happens in a "flash" or a "click". He no longer associates the circumference to the diameter in a certain approximate ratio, but he now sees both circle and diameter as sides of a totality of geometrical objects, generated by the law of the totality, in their proper relationships (e.g., the circle is generated by a point moving equidistant from a fixed point, and the line by a segment moving in its own direction).

It seems clear to me that all questions concerning human experience can make sense only in the light of understanding. Understanding is itself an undefinable. For if I try to define it, then at the end I can ask "Have you understood?" thus showing that you already understood before I started. Indeed, if a person could not understand this, then he would be worse than an idiot, and could hardly be called a human being. The best that can be done is to make explicit a few aspects of understanding.

With regard to experience, this certainly exists on many levels (i.e., verbal, pre-verbal, etc.). But without understanding, each experience is only a fragment. Understanding is the process of coming into being of a totality, in which all is comprehended as a side or aspect of this process of totality. On the other hand, associative thinking is always fragmentary and self-contradictory. We could not hope fully to "verbalize" the action of understanding, any more than talk about light could fully verbalize all that we experience when we see light. However, we may hope that proper use of words can help initiate the action of understanding in the mind of another.

Best regards,
David Bohm

3, Berkshire Road
Bristol 7
Glos., England
February 9, 1962

Dear Biederman:

This is just to let you know that I received your long letter, which I hope to answer in detail when I have enough time to give it the attention it deserves. Meanwhile I hope that you received my other two letters, plus the short letter that I sent you last week.

Right now, I only want to say a few things about the question of "similarity". As I indicated in earlier letters, I do not regard this as an adequate or clear notion, because in some ways, everything is similar to everything else. But how is democracy similar to itself in an essential way, whereas its similarity to fascism, to atomism, etc., is trivial and inessential? In other words, it is necessary to face the question of why certain concepts in Korzybski's scheme are always given the same letter, even though the numerical subscript varies from one case to the other. For example, with democracy, why do we write D_1, D_2, D_3, etc., and not D_1, D_2, F3, D_4, etc., where F is fascism?

I think that this question is a very deep one, and not very easy to solve. First of all, one has to go beyond the opposition of the similar and the dissimilar. Just as there is a sense in which we can understand the totality of circles (as I explained in other letters), so there is a sense in which we can grasp democracy in its totality. This doesn't mean to know all possible details, but rather the essential process that must take place for the generation of a circle, for the generation of democracy, etc. Insofar as we fail to be able to do this, we have not understood the concept of democracy, of circularity, etc. (By generation, I also mean maintenance, and the conditions under which the thing fails to be generated as well.) To the extent that we fail to understand our basic concepts as totalities, our thinking becomes confused and self-contradictory.

I admit that this idea of totality is not yet very clear, but I think it must be clarified if we are to make further progress on these questions. It is clear that the simple idea of identity is false. The idea of totality takes into account the eternal change, the eternally different character of everything. Yet, in some sense, things may be examples of given totalities. I can give you an example in terms of our own feelings. Although each example of anger is always different and changing, isn't there a totality of the process of anger, which is distinct from, and incompatible with, the process of understanding?

Warmest regards,
David Bohm

---

3 Berkshire Road
Bristol 7
Glos., England
March 17, 1962

Dear Biederman:

I have just received the second of your two letters, along with the article by you containing the discussion on Structurism, for which I thank you. Meanwhile, I hope that you have received the copy of the article on "Truth and Understanding in Science" that I have sent you a few weeks ago by surface mail.

Your discussion was very interesting to me. It did not change any of the ideas fundamentally that I had gotten from my correspondence with you, but it was illuminating, and gave me a more rounded view of the problem. The photographs of your work were very helpful. While they didn't give me a good idea of what the geometrical relationships are, they did show that you are using very beautiful colours. I think that it is urgent that Structurist art should use rich, glowing, and "living" colours, as these help give the feeling of an "inner" principle that causes the work to come into being, as it were, out of a force or energy of its own. In other words, it must not appear to be contrived and arranged from an outside pattern by the artist.

When you say that even in mimetic art, the content is not just imposed from the outside (i.e., nature), but is also the artist's meeting with nature, you are stating an important truth. However, this leads one to ask you how you can be so absolutely sure that from now on, photography will be the main way to pursue this line of development. The naive outsider to art tends to wonder whether it is not still possible for the artist to meet nature in new ways, which don't go all the way to Structurism, and which don't give up mimesis altogether. Of course, your analysis has shown that from Mondrian on through Cubism, Picasso, etc., these efforts have broken down and led to chaos. This is certainly an important item of evidence. Yet (both in art and in science), there is always the unexpected. The evidence is never enough to determine the future completely, even in its essential outlines.

So I can see that you are engaged in an important experiment, i.e., to continue the line of development from Monet to Mondrian. You want art to create something new, "existing in its own right", in which mimesis plays no significant part. That is to say, you want the work of art to develop and unfold its own content, without the help of associations to the results of nature's creative process. Rather, it must reflect the creativity in the total process of the world, which includes man as well as nature. That is to say, man is creative in his existence, and reflects nature's creative process too, in his own way, while adding something that belongs to himself alone. Structurist art is then not just a reflection of nature nor even of man's meeting with nature. Rather it should be a unique form of reality in itself, including a reflection both of nature's creative process and of man's peculiar and unique contribution to this process, but also including still more – namely – itself. But then, every creative process adds something new belonging to itself alone. So the distinction of Structurism is that it contains in it the principle that in man's relationship to nature, he can bring new possibilities that are seen and realized by him. Yet, these possibilities shall reflect nature's overall creative process (since the whole of the process is always reflecting into all of its parts and aspects.)

With regard to mathematics and art, I agree with you that the artist should understand mathematics well, and that the scientist (and mathematician) should similarly understand art (and that all must understand music too, of course). There is something of art in science, without a doubt. Whenever a deep scientific or mathematical problem is involved, we have to do with a vision of totality, which inevitably has "beauty", "elegance" in it, in the sense that a deep truth cannot fail to be beautiful, as well as elegant in its pattern and structure. All scientists who have done fundamental work have recognized the great significance of beauty. Moreover, the moment of artistic creation is in many ways similar to that of scientific and mathematical creation. And in art, every beautiful work must be "true" in a certain sense to the creative possibilities of the artist and his medium. In both fields, there can be false appearance of beauty, a "glossy" show that doesn't stand up to deeper understanding. In both fields, what was true to a given stage of development can be false in the next. In both fields, there can be pedestrian work, which is routine and "time saving" aimed more at personal advancement and limited results than at truth and beauty. In both fields, real creativity means putting one's whole being into it, with "all one's head, all one's spirit, and all one's might" (this quotes an ancient Hebrew prayer, but it seems relevant here). In other words, to express totality in your creation, you must be a totality, and not mixed or half-hearted in your motives. Yet, the scientist must also have absorbed into him the large-scale vision of art as well as of mathematics. He comes to his problem with knowledge and with certain instruments, but he must be ready, if necessary, to set it all aside, in favour of something new, if experience shows that there is something false in his starting point. The new vision must then show just how and why the old one was false, and it must contain the truth in the old one. Similarly, it seems to me that the artist could profit by understanding and experimenting with the mathematician's concept of space (although in his final creative moment, this will be only in the background, and would have to be denied and falsified, at least in part). I think that in a similar way, the scientist must understand his own level of experience better, because:

1 All of his data finally come from perceptions and abstractions at this level.
2 His concepts and ways of thinking are subtly influenced by all of this experience.
3 So therefore, both in order to suggest new ideas and to see how he is being unconsciously conditioned by all of his experience, he should understand how the artist is seeing this level of experience.

For example, a better understanding of art would give a scientist a new slant on space, connectedness, relationship, contingency and necessity, totality, etc.; and a better understanding of music could give a similar slant on time process; for example, on how rhythm, counterpoint, etc., combine together to form new wholes. In both these cases, something similar might happen in physics. (But it would be different too, of course.) So there could perhaps be a creative interchange between art, science, math and music, in which each would remain itself, and yet notice the way in which it is also all the others too. But here, the essential point is that in creativity, a pre-given pattern is never what is relevant. Rather, it is an understanding of totality, in which something new emerges. So it would be wrong for the scientist to impose the artistic pattern on his work, as for the artist to impose the mathematician's conception of space in his work, etc.

Now, I shall go on to some of the questions raised in your letter Feb. 24 (leaving till a later letter the answer to yours of Jan. 18). Let us then begin with the question of meanings of abstractions. Thus, we have the series for democracy, D_1, D_2, D_3, etc. And as you say, why don't we include "people's democracy" in this series? You are quite right to point out that each person starts from an implicit totality of meaning. This totality may change from moment to moment, but at each moment, it is a totality, including a very large set of potentialities, as well as the actual cases that are uppermost in the mind of the man concerned at the moment in question. In other words, the term "democracy" calls forth a multi-dimensional structure, with many potentialities, all interwoven into a whole, along with a certain known history, etc. You have called this the basic assumption of the person in question about democracy. I would prefer to call it the basic principle of structure and operation, which is what weaves the potentialities into one whole, and which even more, actually generates the potentialities as sides or aspects of the totality in question, all automatically related by being thus woven together.

Let me illustrate this idea. Most people who use the term "democracy" imply that, in some sense, democracy is superior to other forms of government (e.g., to benevolent despotism). In what way is it superior? Of course, there are superficial superiorities. You cannot trust any despot, and even if he has good intentions, he can be wrong. In a democracy, there is a chance for checking and criticizing such mistakes. But even the people as a whole can be wrong, so that there is no secure guarantee against mistakes. The really essential point is in fact much deeper than this. It is that if the people as a whole cannot find a way to work together freely and creatively, this means that they are childish and immature; so that their potentialities are not being realized. Even if you could assume an indefinite succession of very wise and benevolent despots (which we know is impossible), this would mean that people would be restricted to remaining like little children, taking no part in helping to make their own fate. There is a time to be a child, but then, there is a time to grow up.

So democracy is not only the proper aim of humanity, intrinsic to our real potentialities, but also, it is the proper means to develop them. (Only by actually trying to work together freely can we learn what this kind of work really means.) Here, we part company with the Communists, who agree that democracy is the real end of current social development, but who regard benevolent dictatorship (of the Party) as the appropriate means (at least for the present). And we are even further from fascists, etc., who regard human beings as incapable of anything but being ruled by special individuals or groups who are recognized to be of "superior" qualities.

It is in the light of this basic principle that current ideas on democracy are woven into a totality. Of course, the principle must itself change and develop; and in time, the very idea of an organized system of government may lose its relevance. But for the present, this is very dimly visible in the distant horizon only.

Moreover, the problem of applying these principles leads to endless further developments, even in the present. Consider the fascist and racist point of view. As you point out, this is closely bound up with "identity thinking", which fixes an association of qualities as permanent and inherent. Thus, the racist sees that Negroes are often uneducated, disorderly, etc., and he thinks that the quality of a black skin is tied to a basic quality or principle of being. But how do we see the falsity of this point of view? The answer is that we have enough experience to show that the basic principle is not the colour of the skin, but rather, humanity as a whole. Each of us, raised as the Negro was raised, evidently has the potentiality to act as he has done; while any Negro, raised otherwise, has the potentiality to be different. So the colour of the skin is irrelevant. The basic principle is humanity itself, and its essence is understanding, the perception of truth, and love. To attribute to the Negro a less than human character is to deny truth, understanding and love in oneself and thus to make oneself less than human. And indeed, even some of the animals (elephants, porpoises, chimpanzees, cats and dogs) have this principle to some degree. To fail to recognize this is to destroy a part of one's human potentialities. The problem is to see (with the inward eye) what the truth actually is. Anyone can see that humanity is one in its total potentialities; and he can understand anyone else by seeing the corresponding potentialities within himself. If he excludes the other person, it is a part of himself that he is also excluding. Thus, he creates a contradiction (e.g., if you are angry with another man, you must be angry with his image in you; and the energy of the anger goes on to produce stomach ulcers, high blood pressure, etc.).

With regard to the Communist, the basic fallacy is that one part of humanity can separate itself out and lead and educate another part. This is a contradiction. First of all, the people who are being led are not growing in understanding, but are being conditioned. Secondly, the leaders are also conditioned to believe that they know what is best, so that they are corrupted, artificially separated from others, etc. Besides, there is fear on both sides, which prevents the frank vision of the falsity of the old ways (which inevitably become false as conditions change). People are committed to the old, and because they stand to each other in a relationship of force, authority, worship and compulsion, they are afraid to give up their commitments. So there cannot be truth in such a situation.

You can see then that beneath the concept of democracy, there is an active, growing, changing, "living" principle, which explicates and realizes itself in any number of specific potentialities. It is this principle that is implicit in our minds. We can never explicate it fully. So the meaning is always more than the words and illustrations.

The concept of democracy, for example, has much more to it. Thus, consider love, understanding and truth as principles, which are really what democracy aims to realize, and which are presupposed if democracy is to work. To realize these, we must be ready to see the whole truth about each man, and to try to help him realize his potentialities, whatever they are. Arbitrary presupposition about the relative value of different potentialities may have to be revised. Even the animals have certain potentialities that we don't have; and perhaps to be fully ourselves, we should have to help the animals, in certain ways, to be fully themselves. I sometimes feel that our notion of being superior to animals is somehow interfering with our perception of important truths. It is clear that we are different and in very important ways; but to establish a scale of superiority is false to our essential principle of understanding, truth and love.

It is clear that in explicating the meaning of democracy we pass over into our other principles. So in this regard, democracy is similar to other aspects of the world, as you suggest. But I do not understand what you are driving at when you say that democracy is always different to itself and never similar to itself. Is it not both similar to and different to itself?

Here, there is an important problem. Are we not smuggling in the concept of identity by the back door, when we use the term "self"? Granted that this term should not refer to identity, then to what should it refer? Is there not a real problem here, that requires careful thinking and a new understanding?

A few things can be said now. Thus, democracy is, at each moment, and for each person, an implicit totality. It is changing from moment to moment, and from person to person. Even in a single moment, it has endlessly different potentialities. Should we then say that "self" means a true totality, that can be known only by implication? Even if we accept this, I don't see why a totality is only different to itself and not similar to itself.

In this connection, fascist and racist thinking is not so different from totality thinking as all that. Rather, I would say that the racist, for example, is confused about totality, and not that he thinks only in terms of abstract identities. Thus, the Nazis often spoke of a principle of "Germanness", of "Jewishness", "Negro", etc. This principle was regarded as transmitted by heredity. Their mistake was in narrowly associating the basic principle to secondary or trivial characteristics such as skin colour, hair colour, shape of head, length of nose, etc. But I think that they had the idea of the "Nordic" principle as creative, ever developing, etc. They argued that just as certain genes are needed to make a creature a man and not a monkey, so certain genes make them "Nordic", "Jewish", etc. The falsity is that this was based entirely on fantasy and wishful thinking. But this sense of falsity is superficial, compared with the denial of truth, love and understanding that is implicit in adopting such an attitude. So Nazism (like Communism) is based on false ideas concerning the principle of humanity. It is important for us to understand Nazism, Communism, etc. however, for several reasons:

1 There are ideas which we are all potentially capable of believing if suitably conditioned.
2 We must perceive why they are false, and what is true in them. (Otherwise we are rejecting a part of our own potentialities without understanding, and thus creating a contradiction. Besides we will be incapable of reaching people having these ideas, because we don't fully see why they have them.)

Similarly, with regard to Expressionist art, Surrealism, etc. There is a certain kind of truth in these false ideas; and we must also see, in terms of our own potentialities, what could attract people to adopt such ideas. And in physics, there is a similar stress on techniques and getting "results" that we must understand in terms of our own potentialities too.

So I think a lot of work needs to be done, for a better understanding of the totalities behind our basic concepts (i.e., the implicit totality). About the terms "similar" and "different", I agree with you that to say differences "have no effect in a given field" is too rigid and final. But instead of your term that differences do not "appear", let me propose that when two things are similar in certain respects, then differences do not show or reveal themselves in a certain field. This carries the following implications:

1 In principle, anything can show or reveal itself (by reflection or by imaging) in anything else, but in fact only some things actually do show their differences in certain other things. Those that do not show differences are in this respect similar.
2 "Showing" or "self-revelation" is a basic part of the universal process of the world (it can also be called "reflecting").
3 "Appearance" is a special case of "showing" into the field of our own perception and understanding.
4 We shall take into account your requirement of keeping in mind the areas of nature that surround the particular field under consideration and it still leaves room for "an attitude of tentativeness".

With these ideas in mind, I accept your replacement of "sameness" by "similarity". So now, identity has been dropped from our set of concepts. I would like to propose an argument here for you to think over. Could we not say that "difference" is the logically prior category, and that "similarity" is a special kind of difference; i.e., it is a quality that is different from every kind of difference? This may perhaps seem at first sight to be "word-spinning". But perhaps it means something really important. Thus, we are led to say that there is nothing in the world but difference, if we go into a very basic analysis. But among all these differences, there are some which do not register or show in a certain field. These are the differences that are different from every form of difference that shows in the field in question. Similarly, we could say that in space and time, asymmetry is the basic category, and that symmetry is a special case of asymmetry; viz., an asymmetry that apparently balances out and doesn't show in a certain field. Thus, you can always get symmetry as a special case of asymmetry, but not the other way around. Likewise, you can always get similarity as a special case of difference, but not the other way around.

Generally speaking, our basic categories should be universal in the sense that they contain what are usually called their opposites as special cases. In this way, there need be no contradiction between so-called opposites.

I shall continue this letter later, but I shall send this part off now, as it makes a natural stopping point.

With best regards,
David Bohm

Red Wing
Route 2
Minnesota, USA
March 18, 1962

Dear Bohm:

Received your essay on Truth and Understanding. It is a most beautiful work, and I choose the expression beautiful, deliberately. I think you have broken barriers far more important than that of sound. You have made some discoveries that I am confident will have a major influence in forming future considerations of the problems you take up. Do not misunderstand, I do not want to pose as an authority on these problems you consider. Rather I speak as one who had a most moving experience in reading your essay from the beginning. I experienced a very vivid living quality in what you said, and to that I give the very first credence.

If the "understanding" I experienced in your essay means anything, then assuredly you are on the track of something tremendous. If I do not stop now to express my specific reactions, as I wish I could, it is because I am working day and night to prepare and crate ten of my works for an exhibition devoted to "Constructions." It will take place from May 18 to June 18, at the Stedelijk Museum in Amsterdam. I hope it will be possible for you and Mrs Bohm to see the exhibition.

Your two long letters, dated Dec. 22 and 26, were answered Jan. 18. Your two long letters, dated Dec. 29 and 30, were answered Feb. 24. Hope by now you have both my letters. I also sent you several short airmail letters. It is only in your essay, however, that I began to get the drift of what you were saying about truth and understanding.

Meanwhile, allow me to thank you for sending me your essay. You have permitted me to share in the great pleasure you must have had in being able to say what you did in it. In a few weeks I will be free for re-readings of your essay, and indulge in the pleasure of responding to it, to you.

My very best regards,
Charles Biederman

Red Wing
Route 2
Minnesota, USA
April 14, 1962

Dear Bohm:

I have received your essay on truth, and your letters of March 17 and 23 along with the reprint on scientific methodology. For all of these I thank you very, very much.

Your essay is like something giving motion in many directions to things of paramount interest. I have never read Popper and know nothing about him. But since receiving your essay I noticed two paperbacks by him published in England which I have ordered. One is on Plato and the other on Hegel and Marx, both dealing with the "Enemies of Democracy."

The very beginning of your essay opens up a whole new attitude for me, namely, that there has been a general effort to escape the more difficult, and one might say more hazardous, problem of truth for more tractable goals such as utility, verifiability, operationalism, prediction, etc. Even so could we not recognize that all these efforts were not wholly escapes, at least originally, in that the consequence of these efforts was to establish simple, coherent bases as a *necessary preparation* for deeper searches into the truth? If it can now be said that all these efforts were limited, is it not because of them that we can today go further?

It is a real gain for me to read that no theory can be verified conclusively, *or* falsified conclusively. It is a remarkable discovery giving one a sensation of being aware of a greater totality. Then too, the distinction between associative relating by remembering, and totality relating by "understanding," is extremely important. You are really distinguishing between lower and higher orders of a consciousness of experience. Some 30 years ago Whitehead wrote something pertinent to your discussion. He emphasized the feeling of truth as against its verbalization, that the goal of art is both "truth" and "beauty," and that when there is an absence of the one there is a "triviality" presence of the other.

Associative remembering is largely reflection, whereas, the deeper sensation of truth and understanding is a psychobiological amalgamation of ourselves with the living quality of reality experience. We do not want only to know, but to know in order to be.

You write that "true relations" remain in certain theories even though all theories are eventually demonstrated to possess falsifiability. You cite the theory of gases treated as continuous matter, and that temperature and pressure relations so realized remain approximately true, even though we now know that it is false to assume matter as continuous. I wonder if this paradoxical true–false evaluation of all past searches for the truth, are not the result of either–orism in our evaluation of reality. Are we to regard an apple as a false image of reality because it does not look like atoms? Does a theory pertaining to events on the macro-level possess falsity because it does not apply to the micro- or sub-micro level? Does not the falsity of a theory depend on whether it is properly used, and is not inherently falsifiable in that respect that it continues to possess truth?

Essentially we are concerned with the experience of nature as human experience. On that basis nature is experienced on many related levels, both as continuous and non-continuous. Falsity comes in when we impose the continuous experience as valid for all experience. Just as important, it works in reverse, that is when we impose the discontinuous aspect of nature upon all experience. There is absolutism here in both cases, the latter towards the past, the former towards the future. What I am trying to speak for is what we have been trying to do in our correspondence, to adopt an attitude that seeks the truth in the related variability of nature as totality. To experience discontinuous nature as *actually* becoming continuous in sensory experience. Excellent where you write that it is not experiment that reveals what I would call the falsity potentialities of earlier theories, but a new totality experience that does so. Would you agree, however, that in each new totality view of the universe there remain all the earlier truths? For example, I experience a mingling of form and space in my structural observations of nature which I have the impression is very much like the kind of experience I have read scientists make as a consequence of the joining of time–space. Anyway, there is something else than mere form or space for me. Yet, and this was done spontaneously, I feel a certain coherency to my experience by seeing Euclidean linear notions resting as a sort of first abstraction on these problems, the base of present structural nature experience.

You write that besides seeing the true and false, we see the truth in the false, and the truth about the false. While agreeing, I wonder if this is not to put it falsely. Truth is not *in* the false except when the truth is used falsely, but then ceases to be the truth, there remaining only to know the truth about the false. If the past is limited in imposing itself upon the future, the present is limited in imposing itself upon the past.

You bring up the question of the search for truth by the subjective–relative or absolute–objective attitudes. Sometimes when humans pursue contradictory attitudes over extended periods, there is truth in both of them, which when this is recognized alters both attitudes, bringing us closer to the truth by eliminating either–orism. Subjectivism seems inescapable in some form, and there is no doubt that outside nature is a sort of absolute in the sense only that it is independent of man. We do not doubt nature but only our interpretations of nature, implying that nature is something with definite boundaries to its structure. Yet within these boundaries lie the possibilities for the expression of man's subjectivism to achieve, as with Structurism, what would never appear without man in nature. We seek the truth in order to realize the potentialities of human life which lie dormant in the structure both of nature and man.

Where you write of truth as continual change, that we must be alert to the experience of truth as changing from moment to moment, is very good. But if one must be alert to newer and fuller glimpses of totality, one must be just as alert to new and fuller understanding of all past truths. I believe there is a continuity in man's search for truth, a continuity brought about by the order of his search which inescapably traverses from one related aspect of nature to another. No truth of any time is independent of the truth of any other time, and to the degree we understand this continuity to that degree we can facilitate our continuance of it. Thus the truth in the truth has its own importance aside from the potential falsity of truth. Both permit us to go further.

Perhaps nothing is more important than the emphasis you put on how imperative it is that we experience the truth as in a state of constant change, along with your criticism of the efforts for a "solid foundation." If it is no longer desirable to build the truth like a stone pyramid on the earth, a precise beginning and end, it now being more like navigating in a sea of process, open in all directions, do we not need a ship to navigate the process? And in what way do you propose to build that ship? Will our totality view change constantly, and is that possible, or must we find an orientation, a center of gravity so to speak, that will serve as a coherent base for our response to constant change?

This notion of falsifiability is completely new to me as it is presented in your essay, especially that nothing can be completely false as nothing can completely encompass the truth, has been a very vivid discovery for me. I feel very strongly that it is a very wonderful preparation for the future. It is such a startlingly simple use of the past, so obvious and yet took so long to come by. Recognizing that all theories have a limit beyond which they become false, so now we finally recognize the need of falsifiability as inevitably potential in any theory we make, instead of continuing the futile search for finality.

Let me say that it has been my good fortune to have read your essay. It has made more clear to me the way in which I think I have searched for the truth, and will continue to do so in ways I cannot be aware of now. For that I am in your debt.

You are on the track of something momentous and I wish for you further success in this pursuit, which I am sure you will achieve.

With my best wishes,
Charles Biederman

P.S. If all theories are inevitably falsifiable, what of the theory of falsifiability? If we deny it, we acquire an absolute. If we reject the absolute, we assume it is possible to one day avoid falsification.

P.P.S. Will answer your letters, first chance.

Red Wing
Route 2
Minnesota, USA
April 23, 1962

Dear Bohm:

This is a reply to your letter of March 17. You ask how I can be *so "absolutely sure"* that from now on photography is the proper art of mimeticism. It is difficult to know in which of many ways would be best to respond to your question. You do not take issue with any of the various points I make on this in my published writings, which would give me some clue as to why you are thinking in the way your question implies.

It is not, as you put it again, whether or not it is still possible for the artist to meet nature in new mimetic ways. There is all the art you can possibly want from the past 100 years, that proves the contrary. The question is whether it is usefully desirable, that is, whether new ways are possible which, like past mimetic art, will sustain the evolution of art, beside that of the medium of photography. The answer depends upon our conscious or unconscious assumption about *any* form of art abstracting from the reality of nature. My assumption is that all forms of art that have proved useful both to general human life and to the constant development of art, are those based on achieving and increasing the conscious relations between man and nature. That is, ever increasing the involvement of art in a deeper penetration of nature reality. From the cave artist to the present this has been the crucial avenue by which art has been able to sustain itself as a constantly living experience, by constantly expanding man's experience of nature. In this regard I agree with Whitehead when he says, "Art is the education of nature. Thus, in its broadest sense, art is civilization. For civilization is *nothing other* than the unremitting aim at the major perfections of harmony" (emphasis mine).

What then can we say about all "new" mimetic directed practices in the post-Courbet times? Have not all such efforts, with the exception of *certain* camera-artists, resulted in a heightening not of the interaction of the subjective of man with the objective of nature, but rather attempts to increasingly make the subjective independent of objective nature? This is the opposite of the direction of achievement, not only of all past great mimetic art, but also of all great art which, since Courbet, has left the mimetic effort in the hands of camera-artists. The one attitude expands the action of man's subjective consciousness by means of an expanding experience of objective nature, the other by the futile, because impossible, path of seeking ever greater independence from nature of which the artist himself is a part. Is it any wonder that the latter must depend on the endless arbitrariness of obscurity, waiting for a Godot without knowing if it exists or what it is, if anything?

The confusion of present art is due to misunderstanding each of the steps taken in the gradual change from mimetic to non-mimetic art. This was very, very easy to do because the mimetic element did not disappear at once, and so each advance towards a truly new form of art became, at the hands of the majority of aspiring innovators, easily used for the purposes of some "new" manipulations of the old form of art. I might add, that since the appearance of my *Evolution* book, which re-established the place of nature in any new attempts at art, that since then this has been used by would-be pure creators, some even say they are Structurists, to return to the mimetic manipulation of nature's creations.

On two counts, then, we see evidence of the repeated frustration of those who would assume that mimetic art is not concluded in the old art mediums.

1 None of the major attempts has succeeded in originating its approach to a "new" form of mimeticism. Each has simply manipulated a new achievement in the direction of non-mimeticism, in order to manipulate a mimetic vision of nature.

2 No attempt to continue the past mimetic achievements, that is, to continue ever further penetrations into nature's reality, has been anything but a dry, hollow echo of *what* has been done incomparably better in past art, except in the case of certain camera art.

Only the latter art form has achieved (as did all past mimetic art) a new form of mimeticism, in fact an amazingly new form, and also taken us further into the experience of nature. In regard to the latter, to mention only this one point, the camera reveals what no painter could ever see, let alone paint, what no human could ever see until the camera caught those aspects of the light which escaped vision. The difficulty is that most of us became very habituated to the experience of camera art before we began to realize its immense accomplishments, and how can we "understand" when habit reigns over our vision?

It was a pleasure to read the paragraph where you expressed what you understood Structurism was seeking to achieve. A pleasure because you are the first person to write to me, who does understand. Only one sentence I did not go along with. It's where you say that Structurism is not merely a reflection of nature, agreed, but then you say it is not "even man's meeting with nature." On the basis of the points we do seem to agree upon, how can you avoid seeing that Structurism is still a matter of man's meeting with nature, but on a wider and deeper plateau than previously? The Structurist artist is only seeking to continue the whole process that began with the Aurignacian artist – man's meeting with nature ever anew.

I was very interested in your comments on "beauty" in both science and art. Again I am reminded of Whitehead's remarks which I noted in the previous letter, namely, when experience is at the full, truth and beauty are there together in their fullness, else truth and/or beauty are "trivial." Your Hebrew prayer is well chosen for the point you wish to make between the searcher of pedestrian understanding of truth and beauty, and that which literally envelops the whole being seems so futile to attempt to instil this distinction in others, for it seems to be understood only by those who already understand it in their own life. Such a one does not have to be told for he has no choice but to give his whole heart, his whole spirit and all his might, regardless of social consequences.

I agree that an artist could profit from an understanding of mathematics, even though I know I shall never have the experience. I suspect, however, that the new art must go through a certain amount of development before mathematics can be properly applied, the assumption being that art may demand a unique form of mathematics. So far there have been quite a few artists in the new direction, who appear very gifted and deeply interested in mathematics, and nothing has come of it that could not have been done far better without any mathematics. It would be expecting too much to suppose that present mathematicians are inadvertently occupied with forms of mathematics applicable to art, as it would be to expect the physicists' theories to do so. It is so easy for math to give the artist's vision a sense of certainty which, in fact, leads him away from vision altogether, at a time when a new vision is needed. The new art direction, for that matter, is open to so many false roads of quick results, that some who have taken up my research, in a year or two, speak and behave like old Rembrandts.

It probably is asking too much of you, in view of my illiteracy where numbers are concerned, how the artist can profit from the mathematician's notion of space, but when he comes to his own act of creation he will have to partially deny or falsify the notion. My reactions to such falsifiability rights for the artist's handling of anything science says, is a concession to the artist which the artist I understand would decline.

In regard to the several points you make on the scientist in relation to art, the first certainly belongs first, that the scientist operates on the basis of perception. By becoming aware of the kind of perception that takes place in art, both the mimetic and non-mimetic visions, the scientist would expand the range of his perception, not only into art and that area of nature that comes into focus there, but even in his own field. So far, however, and for a long time now, the easier path has been taken of seeing wherein the perception and creation of the two fields are similar, and this has resulted mostly in crude falsification of scientific theory and a poverty of art theory. What we need now is to discern where the differences lie, which I think will be found resulting from the different aspects of nature with which both are concerned. So far the general effort is to make this whole problem very easy, and which you warn against when you note the artist and scientist must not impose the patterns from their works upon that of the other. Once the best creative scientists and artists begin to seriously comprehend each other, a gross fragmentation of totality could be ended, and an incomparable sense of totality would become possible.

Now to your discussion of democracy. When you change my expression of *basic assumption* to that of *basic principle*, I don't think you change the former but rather go on to a further order of abstraction. When I use basic assumption I mean what all do who speak about democracy, from the most ignorant to all others. You make a further abstraction by pointing out the essential character, the basic principle that is formed by and implicit in the basic assumption.

On your distinction between democracy and other forms of government, and consistent with your analysis, the former offers a structure of human behavior that does not permit absolutes to suppress all opposition to those wishing to enforce the absolutes. The major aspect of this problem, which I am sure you think too, is that the question of democracy is mistakenly identified as wholly a political matter. This attitude has been forming over the centuries out of the efforts of individuals and then with such major activities as art and science.

Your whole essay on truth and understanding is also a dissertation on the operation of democracy. As to your notion, in your letter, that organized government may lose its relevance. Is it organized government, or the imposition of such, that will disappear? Is it not implicit in our present notion of democracy that its potential is to form a self-regulated society, where the majority will become mature enough to see that their well-being depends upon the well-being of others? That is, a world society which, as you make clear, has humanity itself as its basic principle, recognized essentially as the principle of truth and love. The result would be to achieve a new natural state of man, the result of mankind's long journey from that natural state in which nature gave man his birth in the world.

So far moralists, principally of churches, have emphasized that one should do unto others as he would have done unto himself. What you repeatedly emphasize in your letters, the other side of the coin so to speak, that when we hate others we direct hatred against some aspect of ourselves, making ourselves that much less human beings – if this focus could be driven home on a public scale it would do far more than all the ramblings of politicians and priests. I have often tried to explain to some of my well-wishers who advise me not to speak out the truth of art as I see it, to be quiet and thus gain acceptance from authorities, that whatever I do with art to the others, however false their attitude toward art, this I will end by doing in some way to myself and so to my art.

Another excellent point you make, and which the trend of events confirms, is about the Communist basic fallacy, namely, the contradiction that one part of humanity, both in its internal as well as external relations, separates out with the intention of leading all other parts of the world. I agree with your attitude toward the animal world, and we should also include all living forms and the trees, the landscape. We might then avoid too, acting like less than the "lowest" of animals.

I could not find where I said that democracy is always different to itself, never similar to itself. This does not prevent me from taking your to the point criticism, that I have indeed smuggled identity through the back door with my use of the term "self." But don't you think it possible that the notion of democracy may change in the future, to a degree unlike anything we can even imagine today?

Your analysis of the Nazis as confused in their discernment of totality, is very informative. It is worth being expanded into an article, that others might benefit. For if the Nazis and Communists indulge in their brands of wishful thinking toward the world, those in the democracies too often respond to them with just another form of wishful thinking. This is only to emphasize what you rightly insist is the problem, that we must achieve a better understanding of the reality principle, ours and that of others, if we are not to fail in both.

I think the way in which I have been using the term "similarity", that is, to deny identity, sameness, therefore implies that similarities are in fact a special set of differences, and so in accord with the notions you express. We use the term "similar" to ignore particular differences, or to focus on them, as the case may be. These may either be differences that have what we call similarities, or differences that do not have similarities. I say, "The flowers are on the south, the hills on the north, sides of our house." The differences with similarities, flowers or hills, are ignored for the differences between flowers and hills. The differences in nature are like a spectrum of differing differences. The situation here is similar to the case of symmetry and asymmetry, which you also discuss. There is a spectrum of differing asymmetries.

In actuality there is nothing literally similar or literally symmetrical. The former indicates a special aspect of differences, the latter a special aspect of asymmetry. Does this not answer to the necessity you point out for securing categories more universal? In this view there are no opposites to be found in actuality experience. To refer to the actualities in nature which we call similarities, and those we call symmetries, as opposites, it seems to me is like calling the head of a man the opposite of his whole body which includes the head. The issue, it seems to me, is between relations garnered from experience of perception, and relations suggested by meanings words can contain.

I hope soon to answer your final letter of March 23. In the meantime I hope all goes well in your new place and new position. Do you have children?

My best regards,
Charles Biederman

3, Berkshire Road
Bristol 7
Glos., England
March 23, 1962

Dear Biederman:

I am continuing the letter I sent few days ago. I think that I ended on the question of similarity and difference. I said that I agree with you that similarity and difference do not contradict each other, but that similarity is a kind of difference. I wish to add, however, that there are two kinds of similarity.

1 A and B may be similar in the field of C – which is to say that in this field, their difference does not show.
2 But in addition, I would like to propose the idea of similarity for self. As I indicated in my earlier letter, each "self" is a totality of potentialities, generated by a principle of movement and process. Each "self" determines a "natural" field of abstraction, in which there is essential difference and essential similarity.
3 I think that it is necessary to introduce the idea of essential difference + similarity, because without it, the field is too broad. Each "self" determines its essential field. Besides, there are differences and similarities that are inessential or trivial for a given "self" (but which may be essential for another "self"). For example, with democracy, certain processes are essential, especially those having to do with freedom, self-determination of the potentialities of people, people working together in freedom without fear, through understanding, love and respect for truth. Others are not essential (e.g., whether they should organize through local governments, or by their jobs and common interests – this being decided by the merits of each case, with regard to how each proposal serves the basic principle of democracy).

You will notice that it was not possible to define the essential field of democracy without freedom, understanding, truth, love, and creativity (as the realization of human potentialities). I think that you will find that all human activity presupposes these in the background. If this background is confused, everything else will go bad. I sent you a paper on "Truth and Understanding in Science" (by surface mail), and if you don't have it already, you will soon have it. In there, I explained in more detail some of my ideas on the subject. Now, I would like to add a few points. (I suggest that you read the rest of this letter after you get the paper on "Truth and Understanding".)

First of all, in the moment of understanding something, an action takes place. This is, as I pointed out, the act of perceiving a totality. Now, I regard all perception, both inner and outer, as essentially similar. We may say that we become aware of the world through the series: sensitivity, perception, repetition of perception under new conditions, reflection on perception, perception of falsity in earlier perceptions, understanding, application of what has been understood, generalization of the understood, and eventual demonstration of falsity in this generalization, to be followed by a new process, which must take the truth and falsity of the old one into account; and go farther besides. But here, I would like to stress the difference between understanding and what has been understood. Just as you once said about creativity in art, when you understand a given thing, you will never understand it again in the same way. The next moment, you may remember it as the understood, or you may understand it again, but in a new light. This stresses that understanding is an act. I say, it is an act of perception in which the inner, the outer, and their relationship, are perceived as a totality. In this connection, there is a statue of Athene by Phidias (or rather a copy) in the Louvre in Paris, showing one eye of the goddess focused outward and the other inward. This shows that even some of the ancient Greeks understood that true wisdom is the perception of the inner and the outer together.

Now, the problem of the unity of the inner and outer is really the key one in science, philosophy, art and in human relations. Indeed, I would say that the survival of our civilization depends on whether we can really understand this point – that is – not just as a formula in words, but rather, by a real act of perception, in which we can truly say: "I see it".

It is interesting that in every way of looking at the problem, we find logically that we cannot avoid the conclusion that there is this unity. Thus, in physics, we know that each element (e.g., atom) in the world is connected to others by indivisible quanta, which implies that its very being blends and merges with that of its neighbours. With regard to ourselves, where would we draw the line between the "self" (the Ego) and the "non-self"? It cannot be drawn anywhere in the body or the nervous system or the brain. And according to the quantum theory, our bodies link, blend and merge with the rest of the world through forces of interaction, light rays, etc.

If we go to the opposite extreme and just consider immediate perception, we can also see no division between the "self" and the "non-self". Thus, the field of experiencing is just one totality, in which perception is "going on". At first sight, it seems that there is an "I", or "Ego", that is seeing the world. But a little reflection shows that this must be false. In reality, the "I" is perceived in the same field of experiencing in which is perceived the other person, the table, the sky, the rain, etc. A certain part of the whole field of experiencing is labelled "I". This part is associated with certain memories, certain desires, certain knowledge, certain possessions, certain skills, acquired and inherited, etc. The separation of this part of the total field is largely a habit, which is of some practical utility. However, we have somehow gotten the idea that this "I" actually perceives the rest of the world and makes decisions as to what should be done. In reality, there is the individual, in which there "shows" both the "I" and the "non-I" – that is – the whole world. There happen to be different functional mechanisms associated with how this "showing" takes place. Thus, through the eyes, ears and sense of touch, it is mostly the "outer" world that "shows", while through the sensations of pain, pleasure, desire, fear, anger, etc., the "inner" world "shows" in the total field of perception (memories and thoughts are also regarded as belonging mainly to this "inner" world). But science tells us plainly that there is only one world, in which our bodies exist, along with our nerves, brains, etc. And our immediate perception with equal plainness shows us only one world – in which everything is perceived. Our habit of perceiving two worlds (inner and outer) is very strong, but a little reflection shows how false this idea is. Thus, if we get angry at someone then (as I pointed out in my previous letter), this anger is directed at "the other person in me"; so that it goes on into my stomach to make ulcers, etc. Even if it doesn't do this, it directly creates a turmoil that confuses all thought and perception (because I become confused regarding the "other person in me" as being entirely outside of me).

The situation can be put like this: all perception goes by way of what may be called an "inner show" belonging to the individual.

However, it is only apparently "inner", in reality, it is neither inner nor outer, but it is just "the show", in which the whole world – "self" and "non-self" – is revealed. This perception includes thoughts, feelings, etc.; and on a higher level, it includes understanding and perception of truth and falsity. Indeed, because there is no fundamental distinction between "inner" and "outer", there can be no basic distinction either between ordinary perception and understanding, although the functional mechanisms involved in both are very different. Nevertheless, one passes into the other, blending and merging in a whole process.

"The show" goes on in response to the whole world. We understand that there must be more in the world than there is in "the show". This "more" includes not only the outer world around us, but the inner world that we sense by feelings, memories, desires, etc. Science tells us that these two worlds are one, and our logical analysis of direct perception tells us the same. But can we actually see them as one?

I suggest that a child probably does see them as one, and that as he grows up, he rather painfully learns how to split them in two, in order to protect himself. To see the world as one puts a person into a very vulnerable situation, as I shall try to explain later.

I shall try to approach this problem by an indirect route. In the paper on "Truth and Understanding" that I sent you, I suggested that to grasp the basic principle of truth itself in a real act of understanding (and not just in words), one would need a very different kind and level of understanding. Now, one may ask, "What is preventing us from doing this right now?" I think that the answer is that our minds are clouded, because they are full of false ideas, not just about science, philosophy and art, but also concerning everyday life, motives, etc. Now, truth cannot restrict itself to any field of specialization. The whole truth passes from one field to another, first because different fields are similar, and secondly, because everything is related.

If there is a false idea in one field, which is operative and effective there, then it interferes with the free functioning of the mind in other similar and related fields. The field of motivation is particularly significant here. Consider, for example, a hypothetical scientist who said that he understood that a certain theory was false, but who felt that he would like to go on working on this theory, because he had many commitments (e.g., he had hired many people to work on the theory, and if they stopped, they would lose their jobs, etc.). We would say that this scientist had somehow missed the whole point of scientific research, which is that its essence is truth, so that to understand truth and to fail to allow it to operate is an utter absurdity – if not an insanity. Of course, such motivations can and do operate subliminally and subconsciously, to prevent a scientist from pursuing lines that could embarrass him in some way. Nevertheless, once a true idea comes out, then any sane man sees that he must allow it to operate. Thus, at least in science, we can all see that we have not perceived the whole truth unless we have also perceived that the truth must operate. If we think it is the "Ego" or the "I" that operates, we are confused. (For example, that we follow truth because we are "honest", so that the Ego makes a "choice" – as if the Ego could with meaning and sense choose to be "dishonest" and thus to follow a falsehood.) In reality, it is the truth that operates, outside of the preferences of the Ego. And indeed, the truth can even operate on the Ego, by perceiving and understanding its motivations deeply. So what happens is that the basic principle of the individual ceases to be the Ego and is truth instead.

Now, at present, this happens in a restricted domain, such as science or art. But to see the basic principle of truth itself, it would be necessary for the individual to allow truth to operate unhindered in every field. A basic part of the whole truth is to perceive the falsity of every operative idea that is really false. This is extraordinarily difficult, as our motivations are confused and twisted in a very complicated way. Many of our false ideas operate subliminally, or even subconsciously. The problem is far more difficult to understand, than for example the theory of relativity, so that it requires a sustained and serious effort. Yet many people expect to understand truth in five minutes.

Another aspect of the whole truth must be the perception that there is no fundamental difference of "inner" and "outer". One must see that human feelings, aches, desires, ambitions, fears, etc., are no more important in a fundamental sense than are those of other people, and that all of these "inner" workings are just "going on" in the same way that it rains and the sun shines. This is also very hard, but if one doesn't see it, then one is confused, and can only perceive truth in its fragmentary form, and not as a totality.

I have tried working at the latter problem, and after some work, I occasionally got a sudden "glimpse" in which one felt that reality is in a different dimension (as two views of an object in a stereoscope fuse on to three dimensions). In this new set of dimensions, one saw that the inner and the outer are basically one. However, this glimpse lasted for only a moment. I think that I saw why it didn't last. In this state of unity of "inner" and "outer", the new truth starts to operate. But this operation implies a totally different kind of action – an "openness" that is at variance with all the norms of common life. It also makes one very vulnerable, as nothing can be kept for oneself or concealed. To continue in such a state would require a kind of love that does not exist in me, and that probably exists in very few people. So fundamentally, our understanding is limited by the absence of love. This is what I indicated in an earlier letter. Understanding without love is impossible, as is also love without understanding.

In your letter of Jan. 18, you indicated that love by itself might interfere with understanding. But I think that this would not be real love. Rather, it would be attraction and identification. For example, a parent might fail to see the weaknesses of his child. This is because he is identified with his child, and therefore doesn't want to see these weaknesses. This so-called love is then directed primarily at the self, which obtains satisfaction through relationships with the so-called "loved one". But love in its purity does not depend on any satisfactions or identifications for its existence. If it does, it cannot be love, but is a kind of exchange. Such an exchange can lead to hate and frustration when the expected satisfactions are not forthcoming. Thus, we come to the mystery of the "love–hate" relationship, as suggested by Freud. But it seems to me that there is no mystery. It wasn't love in the first place, so that it was only natural that it could turn to hate.

Pure love and understanding would have to exist together always. There would be no motive for clouding the truth, and there would be natural attentiveness which would lead the person continually to understand afresh, if what he really felt was just love, and not something else besides.

Of course, such love is rare in society, but this is perhaps, and in all probability, the result of our conditioning. What actually exists in general is partially love and partially something else, which is mistaken for love. I think that if there were pure love, then the understanding would be unclouded, and we could perceive truth whole (as a totality).

In this connection, I once told you about a painting by Rouault, *The Old Clown*. I saw the foreground and the background, related by complementary colours. Then, I saw the colours pulsating, and suddenly, the division between "inner" and "outer" disappeared. I felt and saw the whole being of the Old Clown pouring outwards, and I felt the whole world including myself to be flowing inwards into him. This was so striking that it gave me pause for reflection. Now, I think that

I understand it better. I think that the painter really saw the oneness of "inner" and "outer" and put it down in paint (I think that when we live this way, the world will be very different, and all our present problems will dissolve). Of course, I may be wrong, but the experience was striking enough to take it very seriously. I can also recall that there seemed to be a tremendous love coming out of the Clown, also a sense of great pain due to his state of vulnerability. Maybe it was just due to my conditioning, as you once suggested (in terms of "resonance" in physics). But maybe it wasn't. Surely, it is important (at least for me) to inquire into the truth of the matter.

This is an example of why I think there may be room for something new, even in the field of mimesis (aside from what can be done with a camera). So as I said in my last letter, I think you are carrying out a very important experiment in the field of art, with implications reaching into every field of human understanding. Yet, your direction may be only one of a number of possible ones, which could and should be followed. What do you think of this idea?

In this connection, it seems that a possible valid direction for the artist to explore is the problem of the unity of "inner" and "outer", of subject and object. In physics, this has become an important problem. Thus, because of the indivisible, quantal nature of all process, an observation is impossible in which the observer does not merge and blend with the object observed, and participate in its very being. On the large scale, apparatus is usually too insensitive to reveal this fact. (So it does not show up.) But with very sensitive apparatus, it does show up. Nevertheless, our usual idea of separating the subject and object prevents us from understanding the indivisible character of all reality. Perhaps if our own minds became more sensitive, they could show that the subject and object are really one, so that we could see that there is a totality in which both "I" and "non-I" are going on, in an essential relationship and interpenetration of being. Then, one might understand the similar problem in physics in a new way.

Is there some way in which Structurist art faces this problem? It seems to me that there is, in the sense that each structure seems to blend and merge with surrounding nature, and is not just isolated inside a frame. Yet, the problem is much deeper than this. The observer should be led, in my opinion, to feel the structure to be moving outwards to the world, while the world, along with the observer, is carried into the structure. In other words, in a real masterpiece of creative art, both the creator and the viewer should have moments of understanding, in which that work becomes the very essence and principle of their being (as happens with operative truth).

Finally, I should like to consider again the question of understanding by a blind man, which we once discussed before. Is it not possible that a blind man will develop new kinds of sensitivity, and in this way, new modes of perception? For example, I heard on the radio an interview with a man who had become blind. He said that he could tell people's character far better from intonations of voice than he used to be able to do by looking at them, because they were able to hide their real -intentions from the eyes, but not from the way they talked. I could not imagine a blind Leonardo or Newton or Beethoven, to be sure. But then, there might be some other side of the totality of all existence that such a man might understand very well, especially if he were helped by those with sight. This seems to me to be an open question, about which there is no need to draw conclusions, without further inquiry and experience. Yet, it raises an important point with regard to the nature of understanding. Is there not a kind of understanding that goes even beyond vision, and may it not be important for us to understand this?

With regard to pre-verbal experience and abstraction, I would say that what is important here is something you once said – "All perception is abstraction". Indeed, perception is the "showing" of the world in the field of experiencing. As such, it is necessarily abstract, since some aspects show strongly, others weakly, and others not at all. So even pre-verbal perception is abstract, in a certain sense. What intelligence and understanding add to perception is of course a key problem that we must study. Here I would like to stress one point – it helps to arrange abstractions according to value.

The primary case of value is truth. Indeed, the words "true" and "valid" are almost interchangeable, suggesting that truth is in fact a value, and indeed, the basic value, or the value of all values.

But there are many other aspects of value. Thus, abstraction must proceed by a perception of what is essential and inessential in a given field. We abstract the essential and set the inessential aside. Indeed, without this ability to see the essential and the inessential, our power of abstraction would not get us very far. For then, abstraction would be only accidental (e.g., dependent on how sensitive we happened to be to each aspect of experience).

I think it is a mistake to say that the more primitive experience is necessarily the more concrete. It may in fact be more abstract (and indeed, it usually is). However, primitive abstractions are likely to contain a great deal of accidental wealth of detail. You are probably familiar with Lamb's essay on roast pig. In ancient China a peasant's house once burned down, and since he kept his pigs in his house, the pig was roasted. On touching the pig, he burned his finger, put it in his mouth, and noted the delicious taste. Thereafter, houses began to burn down with great regularity, until a great genius noted that roast pig could be made without burning down a house. So you see that what this genius did was to abstract the essential from the inessential. The inability of "common sense" thinking to do this adequately is so widespread that I should like to give it a label: "The roast pig fallacy". Hereafter, I shall refer to it by this name.

So a great deal of the function of intelligence and understanding is to assign the right values (essential, inessential, true, false, etc.) to various aspects of experience. As a result, perception with understanding is more concrete than immediate perception usually is, because it adds (among other things) a set of right values, which is at the foundation of intelligent abstraction. These right values are as necessary for the perception of reality as is the original sensitivity, by which we become aware of the world. These right values are not "judged" (i.e., by choice), as a judge would decide whether a criminal is guilty or not. Rather, they are perceived, but by inner perception and sensitivity. Thus, perception of value is as much a part of total perception as is immediate sensory experience. If (with Athene) we turn one eye outward and the other inward, then our vision is far more concrete than it is if we look only outwardly or only inwardly.

So it seems that what is often called the more abstract is in reality the more concrete. Concreteness has more truth in it than accidentally abstracted ideas and perceptions. The concrete is closely related to your idea of "natural abstractions". For what do you mean by "natural"? Clearly, you must be referring to an abstraction containing a true perception of what is essential in a given situation. Such a perception is more concrete than one with a wealth of detail, in which the relevance of all this detail is not properly valued.

Finally, I would stress that truth is concrete. That is, it is a kind of abstraction in which all aspects are properly organized, ordered, developed and valued. All of these additions are what help to make it concrete. Truth contains potentiality as well as what is immediately actual – a totality of potentialities according to an actual principle of movement and process. In this way, it is also more concrete than immediate perception. It is the inner perception related to the outer that makes truth concrete, but changing from moment to moment; as I indicated in the article that I sent you.

I shall continue this in a third letter later, directed at answering your letter of Jan. 18.

Best regards,
David Bohm

P.S. I just received your letter of March 18, for which I thank you very much. (The one on Truth and Understanding.)

Red Wing
Route 2
Minnesota, USA
April 27, 1962

Dear Bohm:

In reply to yours of March 23.

In point (1) you say, A and B may be similar in field C, in which field their differences "do not show." In some fields, however, the differences of the similarities show clearly, in others we know they exist even if they did not show. Wouldn't it be more correct to say, we are ignoring such differences in the particular abstractions that concern us? Wouldn't that make our thinking, our consciousness of abstracting, more in accord with the actuality we assume to exist? Consciousness of "ignoring" implies the conditional status of our use of the term "similarities," which is to be in accord with the basic nature of the inquiry you have been making in this problem.

Point (2), your notion of similarity in respect to "self." I am bothered by the term "self." I have the impression that your use is flavored with some identity. I want to substitute your other term "field," "similarity for field." The latter suggests an order of abstractions that seems important. Namely, if the field is man, then the first order of abstraction would deal with the "similarities for humanity," next for the individual field. This would counter the prevailing tendency to start with the self, and then stay with the self no matter how far and wide we spread the operation of the self.

Your observations of "essential differences" and "essential similarities," strike me as very important. In other words, you are talking about the multi-ordinal structure, the order of abstraction, for similarities and differences, according to the field or fields considered. Your example of democracy gives a good illustration. That is, the general field is first considered, for it contains the basic principles of democracy. As you say, if this background is confused, confusion follows from it.

Your emphasis upon understanding as an act, specifically an act of perception, is exceedingly important. It is not too often understood that this is so for others, as much as for artists, that it is not just a verbal act. As you put it, one must be able to truly say, "I see it."

Your distinction between understanding as an act of perception, and what *has been understood*, is important. We could distinguish between recollected perception and direct perception. As an example you recall my remarks about understanding in art which could be put this way: Every act of understanding is unique and can only be repeated by recollection and never again by direct perception. But there is something unique about each act of perception that passes into the field of recollection, it can be subjected to *direct* perception. This means that each recollected perception can thus be expanded as informative experience for our efforts in direct perception. All works, of any artist good or bad, are always changing, if for no other reason than that every artist is changing.

All the artist's works change with each new perceptive act of understanding from new experiences in nature and from each new work. Failure in these matters means the artist loses his capacity to judge his own efforts.

The way in which you explain that the inner and outer world become a unity in a single field of totality, is a refreshing outlook on an ancient problem. It is process of relating everything within totality, correcting the too easy mistake of relating a part to a totality that is lacking the sensation of process. In this new discussion about no existing between the two worlds, it was more forceful when you said again that the "I" is perceived in the same field we experience the perception of the other person.

It is all too true that the truth does not depend on the Ego's choice, that the basic principle of the individual ceases to be the Ego, and becomes the truth itself. It is in just such effort that the great works of science and art achieve their birth. But, as you say, to achieve this an individual must pursue all the truth that operates in every field. Today this is impossible. The biggest gulfs of all lie between the most important fields. But if we are aware of the truth as the basic human principle, above all that it is not dependent upon our Ego, we leave passage ways open to all the fields around, especially around the field we work the most in, making ourselves accessible.

I was extremely interested in your "sudden glimpses" of another dimension of reality, brought on by your wish to see the totality which is usually separated into parts. I wish you could hand me one of those glimpses so that I could compare it with mine before nature. The larger aspect of each of our perceptions is built on different kinds, so I wonder if what we see is like different sides of a single form. Why should not the totality be such that it is experienced in many ways, just as many persons will experience a particular person differently? The day artists, scientists, etc., can begin to relate their different sensations of totality, totality might begin to take on a coherent, perceptible form.

The search for a literally perceptible sensation of totality has been eternal with man. But if at this point the powers of man to abstract are tremendous compared with the earliest of men who sought to understand *all* the world about them, still we have lost something that earliest man had as a natural endowment. There exists a deep split between man and nature, especially sensory nature, which has been long in the making but is now sharply to the fore if one wants to look at it. It is rampant among both artists and scientists. This split from nature, it seems to me, limits awareness of the truth as the basic human principle, and so limits our sense of being human. It is false, because fragmentary, to seek human principles by focusing on humanity alone. Human life arises out of nature. What an obvious statement! Yet our century has its pompous scientists and artists carrying on like gods who decree the fate of nature. Where you find an art of pure creation that is sterile, thus lacking the living human quality, there you will find an artist confused before nature.

The critical character of the glimpses you describe is that they are non-verbal, the difference between cogitation about and the experience of being totality, a unique order of being biologically totality. Are your experiences such that you can unequivocally state that they were of "another dimension?" I would speculate that animals, birds, etc., have this unity to such a degree that they are not aware of it. The nature of the development of human life has compelled the fragmentizing and categorizing of nature, but as a necessary prelude to becoming aware of the total process of nature and themselves. No doubt man at the beginning had this unity but he became aware of it, and began to inquire into it with worlds of totality: the result – magic, religion, philosophy, science, etc., with art in almost all these different times. I mention this about early man suggesting that it indicates the primary reason we are not able to sustain the sensation of totality. One day man may again regain oneness, another kind of oneness, however, with nature. Surely the common practice of concentrating only on humanity, leads to a fragmentized humanity.

In Structurism the act of creation, which takes place on paper and not by fiddling with pieces of wood or plastic, can only be entered by literally feeling the sense of oneness with nature, really capturing the image of creation prepared for by previous work and experience. Without some sense of this experience of totality, the new art will become, for the artist, or appear for the observer, as barren forms in empty space.

What you call the vulnerability consequences of your glimpses of totality, I have expressed as "it makes me unfit for life," and for the very reasons you state, there resulting a great variance from, even an appearance of absurdity in the eyes of, the so-called norms of life. True enough, to sustain this sensation of totality would require a kind of love, nor can I even remotely claim its possession. But I think you put it all too sharply by concentrating on an ideal goal which demands an impossible achievement of pure love for everything. But a question about the consequences of your pursuit of this goal. Have you not approached it by acquiring degrees of vulnerability?

You criticize my view that love can interfere with understanding, that such is not real love but mere attraction and identification. You cite the example of a parent feeling identification with his child, therefore he doesn't want to see his shortcomings. There is no doubt that your example covers the majority of parents. We seem to arrive at a two-valued situation, if we fall short of its perfection we can only possess an imperfect love that is destructive. If love turns to hate, there can only be one answer, there never was any love. Are the potentialities of human life squeezed into such a narrow area? Perfection is impossible, and we are indeed left only with the possibilities of imperfect love. But I suggest that this is a matter of degrees, of a many-valued situation, that we are not caught between an either–orism, or rather, need not be. There is a range of possibilities from the extreme of false love toward the truth of love. Were this not so, all forms of love, by anyone for anyone or for anything, would remain only false until the day of perfection would miraculously arrive. In all the things we do, whoever we are, we will meet our death imperfect in every one of them. We can, however, if we are fortunate, bring some perfection to some things that we do.

That you had the experience you describe before a particular Rouault is indisputable. All we know is that this particular Expressionism acted as the catalyst. But in what way? Was it merely a stimulus for something that was not there, something that you largely projected into it? For me the experience of this artist is that of an introvert revealing a state of mind about art solely on his Ego terms, an expression that is paper thick. I do not feel the depth of experience comparable to what you experienced as a result of looking at the painting. If you had mentioned instead Titian, Tintoretto, Rembrandt, for instance, any of the self-portraits which Rembrandt did in his old age . . .

I think I already said something about the problem of inner and outer pertaining to Structurist art. Yes, the work must indeed radiate a unity with surrounding nature. Each work is born out of the effort to secure this unity. The work fails to convey this presence of itself within the unity of totality when the artist is impervious to the fact that each work must inevitably enter such totality, or the observer is impervious to such experience in art. I do not think the observer can be "led" into this experience. It should be a mutual meeting between the person and the art, in which the person must bring his share of awareness to the art, as the artist has done to art.

The question of the blind man's perception. I too have heard a blind man on the radio, only last year, make just such a claim. When a man loses his sight he compensates with the remaining major sense of hearing, increasing its sensibility as with his sense of touch. But I suspect that his new form of perception will provide false information, because he can only hear the sound from a person. How often I have first heard then seen, or first seen and then heard, and the impression of the one sense was abruptly discontinued by the experience with the other. We need to see what we hear as a dog needs to smell what he sees.

I agree very much with what you say about abstraction, value, truth. Again truth is the end result, but I note you are not so perfectionist, so either–or, as when you deal with the question of love and its relation to truth. For, if we were just as exacting on the problem of abstracting as you are on that of love, we would have to insist that we must deal with everything our senses abstract because everything is in some way essential to anything we may be doing. To deny this is to imply that the totality has inessential aspects. But we can't approach anything in this complete way. Like everything in nature, like nature itself, we must approach realization by steps along a process, not all, all at once.

You question that the primitive's experience is necessarily more concrete. Let me put it this way: On the visual or pre-verbal level the primitive is more concrete, using this term in the sense that they are more literal in their use of vision. Their capacity to abstract visually is limited, which abstractions quickly enter into absolute verbal abstractions. Your example of roasting pigs by burning the house, exemplifies this. Not until your "great genius" came along was it discerned *in the perceptive* abstractions by discrimination, certain essentials that made it inessential to burn a whole house down in order to roast one pig.

The great tragedy of man is that it is easier to verbalize than to visualize. Man has become overly verbal, thus the rise in our century of so many inquiries into problems of semantic structure as a basic problem of the human race. We have to peel off many layers of words and literature, some of it dating back to primitive times, that coat all that we see. Man has not only increased his capacity to abstract visually but even more verbally, just as he has learned not only to be more constructive but even more destructive. In discussing further your notions on the concrete and abstract, and agreeing with what you intend to mean, I question your means for saying it. I refer particularly to your conclusion that the more abstract is in reality the more concrete. It is my understanding that the character of man's development is an increasing capacity to abstract on ever higher orders, but the danger is in becoming overly verbalistic, words becoming unrelated to the concrete actuality. Hence, abstractions may or may not have the quality of concreteness and, in any case, both types of abstractions are present in what anybody tries to say.

A new correspondent of mine lives in London. He was an art student at the time of the Hungarian revolution, and took part in it. He is now in his early thirties. He refers to Bolyai as "my teacher" and is intensely interested in astronomy. He says he has been a Structurist for two years. I thought you might be interested in meeting him. If so I would suggest to him that he look you up.

Have you seen the interview with Hans Bethe put out by The Center for the Study of Democratic Institutions, in California? It should be called "Portrait of a Harassed Scientist." I hope to read your reprints next, and for which I thank you again.

My best regards,
Charles Biederman

P.S. Yours of April 15 arrived today (28 [April]) as I was finishing this letter, along with the art article, all of which I look forward to reading.

3, Berkshire Road
Bristol 7
Glos., England
April 15, 1962

Dear Biederman:

I am now finishing the answers to your letters of Jan. 18 and Feb. 24.

First of all, we do wish to see your exhibition in Amsterdam. However, we are not sure that we can make it. It is in a middle of a school term, so that it is hard to find time. But the main difficulty will be the money. We have just bought a house and furnished it from scratch, so that we are completely out of money. (A trip to Amsterdam would cost in the order of £75 for the two of us.) However, we will see what happens. Do you ever exhibit in England?

Now to your letters. I think that I have answered them for the most part, but there are a few points outstanding.

1 Creation in art, music, math, etc.

Mathematics is only the outcome, the result of creativity of a certain kind. Just as Rembrandt's paintings will not create any other paintings, so the formulas written in a maths book will not create any new mathematics. All such creation is done by human beings, who must take previous creations into account, but who must also set these aside, and let new creativity operate.

The chief difficulty is desire for security, or rather, the desire to feel secure. People seek formulas, because in this way they obtain a feeling of security. True, the feeling is illusory, and eventually they discover that planned creation is impossible. But for a while, they may feel better; i.e., they may cease to notice the emptiness and nothingness that is at the core of every Ego.

I think that people are confused about creativity. They regard the outcome, the result, the formula or the work of art as the main point, and the moment of creation itself as incidental. The fact is that the moment of creation is the main point and the outcome a by-product, of secondary (but still appreciable) importance.

Our true mode of life would be to live creatively in all of our relationships from moment to moment, and not just in certain moments in art, music, mathematics, science, etc.

Of course, art and science cannot stand alone. Each is in the other. The final criterion of truth cannot be separated from questions of beauty, harmony, love, etc. But the artist also cannot ignore certain questions of fact (e.g., that the human image can be accounted for with the tools of geometry, mathematics, perspective, the laws of light, colour, etc.). It is worth the artist's while to understand space from the scientist's and mathematician's point of view, as the scientist must understand space from the artist's point of view.

2 The Ego

We agree that our society cultivates the competitive Ego or "I" to a destructive extent. The most destructive effect is the way it confuses the mind and the feelings.

Suppose, for example, that I get angry at you. The "you" that I get angry at is in reality the image of you in me. Do I understand you when I am angry at you? Evidently not. So who is the "you" that I am angry at? Evidently something invented by me and projected into my own mind. As long as I reject this "you", I do not understand you as you really are. But if I start to understand you, then I can no longer remain angry at you.

Now, as long as I am angry at the "you" projected by me, a terrible confusion is going on in my own mind. I am really pouring destructive nervous energy into your image in me, and this goes on into my stomach to create ulcers, into my heart to create thrombosis, etc. I cannot fight you without setting up this conflict in myself. So when I am angry, my mind is utterly confused. One part of me (called "I") sets up another part (called "you") and tries to destroy it. But it cannot, because both "you" and "I" are in the same totality of experiencing, belong to a certain individual.

The true reality is the individual, who is alone. But a whole world is in him. He takes a certain part of this world, and calls it "me" or "I", but this is only a habit, initiated by the conditioning of children by society.

A confused mind leads to idiotic behaviour. Thus, if I confuse the window with a door, then I try to walk out of the window, and may hurt myself (as well as other people). Similarly, if I get angry at you, I do not realize that I am fighting myself, and therefore, I also behave in an idiotic way.

Current ideas about "I" or the "Ego" are responsible, in my opinion, for most of the confusion and idiocy that has been characteristic of civilized life for a long time.

3 Contradiction in thinking

On the basis of the above discussion of the Ego, I want to return to a problem that we have argued about before, i.e., the role of contradiction in our thinking. Here, I want to avoid dogmatic assertions, but merely wish to indicate a problem that I urge you to think about seriously.

I can best begin by quoting the mathematicians, who have shown that if you accept a single contradiction anywhere, in any field whatsoever, then you can prove anything you wish, in every field, however distant it may be. Thus, a single contradiction may throw all conclusions open to question.

For example, once when Bertrand Russell was giving a lecture on this point, he was asked to prove from a mathematical contradiction that he was a Pope. He said the following:

> Let us start with the contradiction 3=2. Subtract unity from both sides. We get 2=1. Now, the Pope and I are two. But since 2=1, the Pope and I are one. Therefore, I am the Pope.

So logical reasoning is really a single domain, and you must be careful; for if you accept a contradiction that is too subtle to have been seen, you may make your reasoning invalid, even in apparently unrelated fields.

Now, suppose that you go outside on a clear starry night. Think of the universe, that you are seeing. Think of its immeasurable immensity, and all that is in it. Then think of yourself. Now, as long as you feel that you are a separate Ego, you must say "The Universe and I are two." Yet, you know that this is not so. Physics tells us that our bodies and brains are part of the universe, the totality. Similarly, in our immediate experiencing as individuals, we see only one totality, in which there is experiencing of "me" and "the universe". So we must say "The Universe and I are one."

Thus, there is a contradiction at the heart of all our thinking. This contradiction is accepted by us implicitly in our daily habits of life. Therefore, all of our thinking is confused and open to doubt, likely to lead to idiotic behaviour (as in fact it very often does). Unless one can be in a state in which this contradiction is absent, it is no use for him to say that there is no contradiction.

If you accept a contradiction in any field, then it may lead to an error in every field. And this contradiction about "I" must evidently operate in every field. For every time we isolate anything, we treat it as if it were an "I". The basic pattern of confused thinking, which extends through our whole experience, begins with our ideas about "I", which we learn from our competitive society, when we are very young.

Now, I am proposing an experiment to you. On such a night as I have described, try to say to yourself: "This individual is experiencing; and in this experiencing is a whole universe, a part of which is an 'I' or 'Ego'." If you are really serious about it, you may find that everything is different. The activity of the Ego may subside and take its proper place in the whole. Then the contradiction disappears. Thinking is different. Perhaps in such a state, the individual can think more clearly, and avoid the universal contradiction that I described above.

Clear thinking requires that there be no contradiction anywhere in the mind – and this means that there shall be one individual. We tend to imagine that we can remain the sort of divided beings that we are (dividuals – one may say – instead of individuals) and still think clearly about selected topics such as physics or art. The fact is, however, that as long as we are what we usually are, our minds must be confused. The confusion that arises when we try to relate ourselves to the universe or to another person carries over into physics, mathematics, art, politics, and everything else.

There is a problem of great significance here. All that we can know about the universe is projected into our experiencing by the action of our own minds, in response to our perceptions. How do we come to understand? Does not the possibility of understanding mean that somehow, the notion of separation of experiencing and what is experienced is a confused and self-contradictory idea? If so, can we think clearly on this problem, without first being clear and unconfused in what we are? How can we do this, in spite of society, which is always conditioning us to become ever more confused on the question of the Ego, and through this, on every question that can be asked?

4 Understanding and the understood

I am not sure whether I called attention to the distinction between the act of understanding and what has been understood (the understood).

The act of understanding takes place in a moment. Then with the passage of time, its implications are unfolded – and we have the understood, which is in memory. This is a mixture of truth and falsity. The fact that it is such a mixture must be seen. And then, there is a new act of understanding. Thus, understanding is a very dynamic process. If the mind is sufficiently alert and attentive, understanding is always on the move; and in this movement, one is in a state of true action.

There is truth and falsity in the content of what has been understood; and this must be perceived. But there is no room for falsity in the act of understanding. This must be a true act; or else, it will be confused, and lead to idiotic behaviour.

So we must be in a state of true act. In this state, we perceive the truth and falsity in the content of all our ideas, from moment to moment.

5 Randomness, chance, contingency, etc.

Let us return to this old perennial problem. Thank you for your little note "The Molecule: Order or Disorder?" Let me say that it contains, in germ, the idea that I am aiming to establish in physics. However, we physicists cannot leap quite as far and as fast as you artists; and I shall now try to explain why.

First of all, recall that (as I explained in the article on "Truth and Understanding") an important part of progress in science is to see the truth in the false and the truth about the false. Every idea that we can state in physics will have some false content, but no one would have taken it seriously in the first place if there had not been some true relationships implicit in the idea. So let us try to see the true relationships in physicists' ideas about randomness, as well as the truth about these ideas (the reasons why they are false). This may be an important clue as to what the next step must be.

Let me begin by explaining the physicists' idea of random movement a bit more. Consider a pin-ball machine, such as you have surely seen. I am assuming a rather complicated distribution of obstacles, indicated by shaded areas in the diagram (their precise distribution is, however, not important). Now, imagine a small ball, which follows the dotted line, and which bounces off the obstacles elastically. Its track is indicated by the dotted line in the diagram.

As you can see, the track of such a ball is very complicated. But there are certain features of this track, which are relevant when we consider its overall or average characteristics.

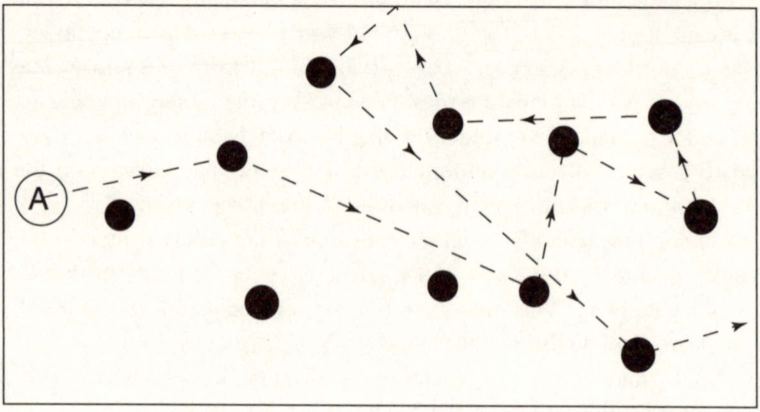

1 The precise nature of the track is very sensitively dependent on the position and direction of motions with which the ball starts out. For example, if the ball had started at a slightly different position from A, it would have struck the first obstacle in a different place; it would have bounced off in a different direction, and its whole track would have been very different. This characteristic is very important: *the details of the track are very sensitive to how it starts.*

2 Now imagine a whole collection of balls coming into the machine, each in a different position. Even if they all start with parallel directions of movement, it is clear that because of the sensitivity described above, each ball would behave in a radically different way. They would soon cease to be moving in parallel, but would soon be in a state where no direction of movement was favoured statistically. In other words, in the long run and on the average, just as many balls would be moving in any one direction as in any other. This is part of what the physicist means by randomness. The initial regular array of parallel movement would be broken, and instead, it would be found that the movement of each individual ball was essentially unrelated to that of any other such ball.

A similar process occurs when a stream of molecules strikes an obstacle. No obstacle is in fact needed. For if they are just slightest bit off parallel in their movement, they will start to collide with each other, and eventually a random array results.

3 If you follow an individual ball, you find that in the long run and on the average, no direction of motion is statistically favoured. Hence its track is said to be a "random walk". The direction after, for example,

the second collision, is essentially unrelated to that after the 452nd collision.

4 We must go more deeply into the meaning of the concept of essential unrelatedness.

Of course, if you were to calculate the track in absolutely precise detail, the track after the 452nd collision would be seen to be determined by the starting track. (We are working on the assumption of classical mechanics, which is completely deterministic.) However, as you yourself admit, there are natural thresholds appropriate to each level. For example, for our large-scale level, effects of quantum mechanics and individual molecular movements in the balls are necessarily below the threshold where it makes sense to try to consider more accurate determinations. In other words, every form of determination and every form of existence must have a natural threshold of sensitivity of response. So it makes no sense to talk of determinations of unlimited accuracy. Yet, you can see that to determine what a ball will do after, for example, a million collisions, may well require an accuracy of determination of the initial position and velocity that goes beyond what is meaningful, in accordance with the natural threshold of our level. So in terms of this natural threshold, the millionth collision is essentially unrelated to the first.

So in the concept of randomness, we have the idea of essential unrelatedness, either of one ball to another, or of the later history of a ball to its earlier history. This makes an important contribution to the understanding of our experience in general, as we in fact do find that frequently, one part is essentially unrelated to another.

5 But now, we must see the limitations of the idea of essential unrelatedness. Here is where your ideas can come in. In a larger totality, it may perhaps correctly be said that there is another kind of relationship. Thus, for the whole collection of balls (as well as for the average behaviour of a single ball in many collisions), there arises a statistical regularity. This regularity is predictable and observable, and agrees with what is, in fact, found, when you do the experiment.

Now, is there a contradiction between randomness on the small scale and statistical regularity on the large? The answer is "Of course not". Each is perfectly compatible with the other.

However, here is where your idea comes in. On the large-scale, long time, and statistical level, new kinds of relationships become relevant. I think that it is wrong for the physicists to try to say that large-scale order is the result of microscopic disorder (or the equivalent, to say that large-scale relatedness is the outcome of microscopic unrelatedness). Rather, one should say that large-scale "order" is a fact of existence, just as real as is the microscopic "disorder". In a still larger totality, which includes both micro- and macro-levels, you are right to say that the molecules behave as they do in order to achieve the large-scale results that they do. However, I think you are wrong to ignore the need to consider the various levels separately as well. For in science, everything is some kind of abstraction. It is necessary to see the world from within that abstraction, to see the truth in the abstraction and the truth about the abstraction (why it is limited in validity and in that way false). Even the abstractions proposed by you will be, in some way, false.

So here is where contingency comes in. Every concept is an abstraction. Within the field of this concept, certain things must be unrelated, or otherwise, there will be no room for newer and higher kinds of relationships in a broader totality. Indeed, man may be said to impose his own relationships on natural processes, and this is possible because various parts of these processes are not already held in iron-bound relationships.

What is unrelatedness in one field is necessary for higher relatedness in another field. So the totality must include the related and the unrelated.

In another letter, I shall have to go into the question of continuity and discontinuity, symmetry and asymmetry, from this general point of view.

With best regards,
David Bohm

Red Wing
Route 2
Minnesota, USA
May 7, 1962

Dear Bohm:

This is a reply to yours of April 15.

I should not blame you and Mrs Bohm for not getting to see the Amsterdam exhibit, since I see that the cost of the trip is certainly not a minor one for you. Perhaps one day I will show in London, and perhaps by then I will have more of my later works to show. So far a number of English artists have helped themselves to my work and writings, more than a little, but none has ever made an effort to arrange for me to publish an article, let alone to exhibit.

Your explanation of mathematics as creation makes clear what you mean. This corresponds with what I mean when I say that mathematics acts as a substitute for creation in art. But then, most alleged creation in art is largely exploitation of what has already been created by others. There is, in short, a fear of risking the demands of true creation. Even Structurism, in the hands of the majority who use it, has been largely exploited in the manner of personalized variations on what has already been created.

So I concur with you when you lead this into a discussion of a "desire to feel secure." The false seeker of creation in art, however, does not feel secure within himself but only in the opinion people of influence take of him. I disagree that the formula is sought as the means for gaining security. No one can escape some semblance of formula, like it or not. Therefore, more fundamental is the distinction in the kind of formula sought, the attitude toward the notion of formula. This in turn would lead to distinctions between different kinds of security that humans seek.

That the general attitude is to regard the created, not the act of creation, as primary, is a good point you make. I haven't thought of it in exactly this way, yet that is just the point that I try to put across in my writing on art. The way you put it, however, hits the essential point of the matter. In my Dialogue you will have noticed my emphasis upon "life as a creative *experience* and *expression*." That is the essential role of Structurist art, to give expression in its purest form to the creative experience which is fundamental to all, from the farmer to the artist. In all our discussions about truth, love, understanding, we should have emphasized that these goals had the essential role of freeing us to be truly and fully creative in all aspects of our lives. In place of what ordinarily leads to anger and ulcers, of which you have written often, one releases the desire to comprehend, freeing ourselves into an act of creation, in place of destruction upon the self and others. Anger tries to separate or fragmentize what is inseparable, the creative act of understanding relates and makes possible a former sense of truth.

This latter effort, however, leads to complications, what you call becoming vulnerable. For then there arise entirely new relations, new problems. But it is just before the sensing of the new that the creative act flourishes. I wonder if you will agree, that to the degree that one becomes vulnerable he also becomes invulnerable, that is, to those actions that fragmentize, confuse, actions that lead to chaos rather than creation. This is the first time I have heard this amazing idea, that if you accept a single contradiction about anything, than you can prove anything, in any field. You note that there is the possibility of accepting some subtle contradiction that escapes us, and thus throws us into falsehoods all down the line of totality. I hope you are not suggesting that it is possible to avoid what seems to me inevitable and inescapable. That there is so very much that need not be inevitable, cannot be denied. But it is as Korzybski once remarked, It's not that we don't know, but that we don't apply what we do know.

Rather than saying logic is a *single domain*, wouldn't it be more to the point to say that it is compelled to operate in the single domain of totality, in that which is not logic, the non-verbal world of nature?

I cannot recall specifically my past experiences with witnessing the night sky when alive with stars, in the context you mention, except feeling the sheer infinity of this seemingly ungraspable universe. But during the period of my more vivid consciousness of nature as a Structurist these last 25 years, I do not recall reactions of the universe and I as one or two. What I have been doing is what you suggest, being an *individual* experiencing nature or the universe as a whole. But in my case I have just done it without giving it any particular attention, except where structural problems of vision enter. Just the same, to see nature as the Structurist does is precisely to gain the fullness of an individual experiencing, rather than limiting experience to a "piciune" (Cézanne) self. John Constable, the English contemporary of Turner, put it well when he said something to the effect that nature shows nothing to the arrogant artist.

Anyway, as a result of your bringing all this up sharply to my consciousness, several nights ago I was walking from our house to my shop late at night when the sky was packed with stars. As I looked up I immediately recalled your discussion and decided to try out your experiment. But almost immediately I forgot my intentions and became engrossed in my usual Structurist experiences. In the midst of this I recalled my original intentions and the following took place. I had the sensation of not being a single form in the sense of boundaries, but of really being a structure of forms and spaces, or forms–spaces, like the experiences I was having of nature at the time. Then I began to cull some abstractions from this, and developed this line of interest. We (humans) are single enclosed forms. We are not like a tree, flower, and so much that literally spreads itself into the surrounding aspects of nature. We are like a piece of sculpture, a single piece all around, all rounded off from nature. A tree's "consciousness" could never form an Ego, but with consciousness humans can go further in what the tree does naturally and spontaneously when it reaches into the earth and towards the sun and rain.

Your distinction between dividuals and individuals is certainly apt, as are the consequences you draw from them. Is not one of the great sources of difficulty here in that man, as he has become more and more urbanized, has increasingly regarded nature as something to conquer, seeing himself as an Ego in competition with a primitive nature around him? And, as man sees nature, so he sees himself whether he knows it or not, and so competition between men, a refined sort of social primitivism. And today nature has dropped its ace on the card table, leaving man the conquerer of nature absurdly contemplating self-destruction.

One thought I forgot to mention regarding your notion of *this individual is experiencing*. When I am aware of myself in relation to nature, it is on the basis of an "amalgamation" (Cézanne) of *man* and nature as I have expressed it before. It is not merely something called "me" experiencing, but a me seeking to experience nature through the experience of man in general, past and present. An effort to absorb the accumulated experience of man as artist as the fullest means potential to me experiencing, in contrast to the "trivial" experience emanating from a confused Ego me. So I would put a phrase at the beginning of your sentence, to read: "Through the accumulated experience of man, 'this individual is experiencing; and in this experience is a whole universe,'" etc. By bringing man into the statement, we avoid what you rightly want to avoid, separation of what is in the experience which leads to confusion, contradiction.

I was very interested in your comments on the distinction between understanding and understood: the former as a flash, the latter as a gradual unfolding always in motion. For in this you say much better what I tried to say in the letter before this one, which you have not yet received. I would argue, however, that it is not only being sufficiently alert of our understanding being in constant motion. This seems inevitable, for nothing can stand still, even our consciousness cannot. It is necessary to be alert as to the nature of the direction of motion which our understanding is taking.

Now for your explanation of randomness, etc. These are perhaps the major problems which modern science has brought to attention which, if brought to some less disputable condition, might end in giving crucial insights into the problem of creation. And I appreciate your patience with my ignorance, in making a further effort to express the notions of randomness, etc. I don't know how many times I reread this part of your letter, trying to understand first what you meant to convey. Yet, each time I failed to understand the justification for the term "randomness." Even so, your exposition was very clear and relieved me, I hope, of some portion of my ignorance.

I'd like to ask: In the experimental field, as distinct from the theoretical, is the notion of randomness essential? I mean, would present experiments become impossible without this notion? For it is beyond my imagination to conceive how a physicist would claim to be conducting experiments directly with that aspect of molecular structure considered "haphazard, without aim or purpose or principle, heedlessly" (Oxford dict.).

You describe the acute determination involved in a ball colliding with a series of obstacles. That no direction of the various movements favors any one direction, statistically, more than some other. This, you say, is part of what physicists mean by randomness. But just what is happening here? Have physicists achieved structural comprehension of a new level of nature? To say something is haphazard sounds more like incomprehension, but which has the advantage of being irrefutable until experimental evidence will discover the character of this new structural aspect of nature. What the physicist has in fact said, is that the atomic level does not respond to statistical order, as does the large scale, therefore there is no order, as though the last word in order had been reached with statistical procedure. One might with equal reason refute the large scale for not displaying the disorder of the small scale. With equal reason people once refused to accept structural descriptions of the molecular level, simply because this seemed so alien to everyday experience.

One thing seems quite certain. If there is a particular kind of order on the discrete level of nature, we will only experience disorder until this order is discovered. If, however, we act on the conviction of disorder, we have surely closed the doors to discovering the order that is to be found. Until the disorder view is proven, the order view must reside at least with equal attention in our perceptive activities.

From the position of randomness you logically infer the problem of essential unrelatedness respecting the individual balls or molecules.

You enforce this conclusion by reference to the natural threshold of determination that is appropriate to each level or field of nature. On the atomic level, you explain, it ceases to make any sense at all to seek determination, as this would involve a fantastic pursuit. Here again we have the same sort of problem described in the preceding two paragraphs. You have not shown that determination is exhausted, you have only shown that a particular sort of determination is exhausted. We don't prove that determination procedure is inappropriate by showing how a *particular form* of determination fails. We only prove that the determinate method of classical mechanics fails, nothing else. Ever since man has been making formulations of the determinate structure of nature, these have followed a course of change in the very notion of determination as more and more of nature was understood.

Just the same I don't want you to think I am arguing against your clear exposition of essential unrelatedness, especially where you refer to abstractions and the contingency factor, for without it, as you say, the possibilities for abstractions of newer totality relations would be impossible. One could also note, that if everything were essentially related, there could not be the kind of world we experience, nor humans to experience. There would be no objects, no different nature levels, in short, no entities of any kind. For nothing is only relations.

I would like to argue a little on your last thought, that man *imposes* his own relationships on natural processes, because aspects of the latter are not held in absolute relationships. I do not think man imposes anything, any more than he imposed the existence of his own kind. Man only discovers the possibilities of introducing new relations which the structure of nature permits. To impose is to compete, and I don't think nature has a competitor in man as yet.

Your letter of April 30 received. In the middle of this, will answer first chance.

With my best regards,
Charles Biederman

3, Berkshire Road
Bristol 7
Glos., England
April 24, 1962

Dear Biederman:

This is just a brief letter to finish up the discussion of your letter of Jan. 18.

You ask whether one shouldn't have a spectrum of descriptions of experience. For example "thinking–feeling and feeling–thinking".

I say that I agree with you on this point, and would like to develop the idea a bit, in relationship to the totality of experiencing, and the understanding of this totality.

Now, just as I made a distinction between understanding and the understood (which latter is the memory or trace of understanding, a mere reflection), I also want to distinguish between experiencing and experience (or what has been experienced). Experiencing is the act itself, while experience is the reflection in memory of experiencing.

There can of course be partial understanding, e.g., intellectual understanding and "feeling understanding" (this latter could be called sympathy). But for a true understanding, the totality must be present. Thought, by itself, is powerless to initiate action. It merely tells us how to reach our goals and the probable results of our actions. The feelings can initiate actions, but by themselves, these cannot in general lead to intelligent actions. So if a person has only an intellectual understanding, his feelings, aims, goals, desires, etc., may be badly confused, so that he gets all mixed up when he acts. If he has only sympathy, then his ideas may be confused, so that he also gets all mixed up when he acts. With a total understanding, he is aware of his feelings, desires, goals, training, conditioning, his perceptions, his thoughts, the end which he has in view, and the means to achieve this end. All of this is integrated into one understanding, which includes also the action that is appropriate in this totality of experiencing. In other words, action is not opposed to thought or to feeling, but all these are present in one totality of experiencing.

At a given moment, one of these may dominate, but the others are still present. Thus, we may have thought with feeling and action in it; feeling with thought and action in it; or action, with thought and feeling in it. But we have always the three: thought–feeling–action. Of the three, action is the most fundamental. For even thought and feeling are actions. They are mainly internal and nervous actions, but still they are actions. So roughly speaking, action divides up in three ways:

1 Physical acting
2 Feeling
3 Thinking

Even perception is a kind of action. It is not merely a passive imprinting of an outer physical world on an inner mental world. Rather, it is an inner show that our whole being puts on in response to what is sensed. Only through understanding is there the possibility that this inner show is true (and not illusion or hallucination, for example). So we may now consider four kinds of actions: perception, feeling, thought, physical action.

In addition, there are further kinds of actions. Thus, when one perceives something, one may desire to possess it, if it is pleasant, or desire to get rid of it, if it is unpleasant. So there is desire, and its movement towards fulfilment. Onc may desire an ideal, invented by society (e.g., religion) or by the individual himself (e.g., ambition to do some great or unusual thing). Desire is generally fixed and limited by conditioning and by the individual's reaction towards or against this conditioning.

There is the totality of memory, including past desire, fulfilment, frustration, likes, dislikes, judgements, conclusions, identifications, etc., etc. Every new experience calls up the whole of this complex. This complex (which can be called memory) also acts. For example, if you have insulted me once, then when I meet you again, I react with the memory of this insult. I do not meet you afresh, and see you as you actually are at this moment, but in my act of perception, I put on the "inner show" of the man who insulted me. Thus, I see you only through the screen of past memories, conditionings, conclusions, etc. In my perception, you are the hateful man who insulted me, and nothing more.

So we must also add to all the other actions the action of memory and the action of desire.

For a comprehensive understanding, it is necessary that all of these actions be in the act of understanding. If a person does not follow all of these, his mind may become confused. Thus, I may meet a man of another race or religion, or nation. Through my conditioning, I do not see him, but put on the "inner show" that he is a typical Negro, Catholic, Communist, Methodist, American, German, or what have you. So I confuse actuality with a reaction of my memory as conditioning. Such confusion must, of course, lead to idiotic behaviour (which has been characteristic of men since the beginning of recorded time, if not before). Thus, if I confuse a window with a door (because they are both holes in the wall), I may step out of the window and hurt myself and other people. If I confuse an individual man with a Catholic, Negro, American, etc., I become equally mixed up, and my behaviour is likely to be equally idiotic.

So we also have the action of confusion.

Of course, in the act of understanding, the whole being is integrated, so that our habitual state of confusion tends to die out; and thus, it ceases to act.

Our confusion is in every field (thought, feeling, perception, desire, etc.).

A large part of our mental confusion comes from a misunderstanding of negation. We usually think only of exclusive negation. Thus, I can say "A is definitely not B", meaning by this that A excludes B. But if I say "A is not definitely B", I say that while A and B are different, they do not exclude each other.

In the act of understanding, we have non-exclusive negation. At the moment of complete understanding, I negate all my past knowledge, not by annihilating it, but by comprehending it as a side of a new totality. In this way, I retain the valid features of past knowledge, while dropping those that are false. So I could say that understanding is comprehensive (i.e., non-exclusive negation).

The essence of process or act is comprehensive negation. Thus, if a person says "I am walking", this implies an "I" or an "Ego" who is the subject, and who initiates the act of walking. But if one says "I am in the act of walking", this means that the act is first. This act contains (or comprehends) both "I" and the "walking". It is not initiated by a subject, but rather, it generates a subject as a side or aspect of itself.

Similarly, if one says "A is becoming B", this implies that first there is an entity, A, which initiates a process and thus becomes B. But if one says "A is in the process of becoming B", this implies that the process is first in a logical sense, and that it contains a past side, A, and a future side, B, both being generated, as it were, in the process itself.

So we should regard the verb as the basic form of language, while nouns are aspects of verbs. At present we tend to regard the noun as basic (the subject), and all actions or processes are thought of as originating in "things", as represented by nouns (which are the "subjects" of verbs).

This must be done in mathematics too. The mathematical symbols must refer to actions, and processes, and not to things. Things are to be referred to as sides of actions and processes. I am working on this idea now, and have made a bit of progress.

So understanding is an act in which the totality of experiencing is comprehended as sides or aspects. In each new moment, the past is comprehensively negated as a side of a new and fresh understanding (or at least it should be). When we do not live this way, we try to repeat the past, and inevitably become confused, since we regard the understood as capable of being completely true. In fact, only understanding can be completely true. The understood is always a mixture of truth and falsity.

We see then that each act of understanding must comprehend a whole. If at a given moment, the intellectual side dominates or is emphasized, then it must comprehend perceptions, feeling, physical action, etc. When the feeling side is emphasized, it must comprehend the intellectual side, as well as all the others. Even when we are doing something physically, this act must comprehend all the perception, thought and feeling, that had gone into it. Likewise, in perception, there must be a comprehension of thought, feeling and physical action, which contribute by entering into the "inner show" with the aid of which we see the world.

In this point of view, the separation of "subject" and "object" is dropped. What is fundamental is the act or process itself. "Subject" and "object" are non-exclusively comprehended or contained as sides or aspects of the act or process as a whole. Each act or process emphasizes one aspect, but nevertheless, all aspects are contained in each. Thus, as you suggest, we can have "feeling–thinking" and "thinking–feeling" with a whole spectrum of possibilities. Only it needs to be extended to: perception–feeling–thinking–physical action.

Even this only hits the main points. There are many finer gradations. Besides, there is the aspect of confusion, which is now habitual, mainly because of our training and conditioning. Our minds are generally confused, and only occasionally are they clear. What is needed is that they should generally be clear, and only occasionally confused.

My wife and I send you our best regards. We are a bit optimistic now and hope that we can get to see your work in Amsterdam, but are not yet certain.

Yours,
David Bohm

CHAPTER SUMMARIES

Paavo Pylkkänen

CHAPTER I: A NEW VISION OF TOTALITY

March 6, 1960

Biederman writes to Bohm for the first time as a result of reading Bohm's 1957 book *Causality and Chance in Modern Physics*. He sees the development in physics and art to be analogous: just as quantum physicists responded to the crisis in classical deterministic physics by postulating indeterminism as fundamental, so, when the old view of nature was no longer adequate for art, the Surrealists, "unable to wrest a new order from nature, assumed the view of nature as 'disorder'". Biederman, a seeker of a new order in nature, is clearly fascinated by Bohm's argument in *Causality and Chance*, which opens up the possibility that even the quantum level can be given a deterministic description. Biederman says that Bohm is approaching the same problems as he himself, but from another angle, and emphasizes the need for a dialogue between art and science.

March 26, 1960

Bohm replies by agreeing about the necessity of bridging the gap between science and "other fields of human endeavour", as these have a strong effect upon one another in any case. For example, he points out that Niels Bohr's interpretation of the quantum theory is related to Existentialism and Positivism in philosophy and is a part of the wider "Modern Movement", which includes radical developments in art, literature, etc. Bohm says that there is a kernel of truth in these new developments as well as a great deal of falsehood. He ends by calling for "a new overall point of view".

April 11, 1960

Biederman raises a question which becomes one of the recurring themes in the correspondence: "How does one distinguish disorder from order in the structure of nature?" He then asks for a further clarification about the influence of other fields on Niels Bohr's ideas about physics. He is concerned about the adverse effects science has had on art and similarly admits that art has failed to benefit science. He then proceeds to tell Bohm about his own work, which can be seen as a new development in art – both his own art and his theoretical descriptions of the new approach in his monumental 1948 book *Art as the Evolution of Visual Knowledge* and in subsequent texts. Biederman emphasizes that, contrary to Bohm, he can see no kernel of truth in certain developments in art, such as Surrealism, Expressionism and Action Painting. He finally mentions how his latest book *The New Cézanne* (1958) tries to show "the transition from the old mimetic view of nature to the new view of nature as a creative process".

April 24, 1960

This is the first long letter in the correspondence. Bohm begins by discussing the relationship between Bohr's quantum philosophy and Existentialism. In quantum physics there is an indivisible link between the subject (or his proxy, the observing apparatus) and the object. More generally this means that each part of the universe is indivisibly related to the whole and that all parts, considered separately, are *ambiguous* in their mode of being. Existentialism, on the other hand, emphasizes a similar ambiguity of parts in the psychological domain: our existence as separate psychological entities is ambiguous. This underlines wholeness, but Bohm wants to go beyond Bohr in trying to formulate a mathematical and physical theory of the new relation of the part and the whole, a kind of "vision of totality". He then goes on to outline this new point of view. The totality of the universe is infinite in space and eternal in time, but we introduce various basic distinctions: part and whole, past and future, etc. This leads into a discussion of freedom and necessity, and Bohm tries hard to formulate a view in which the present moment is not entirely determined by the past, a fascinating theme which develops in many of the later letters. He makes here an analogy with art: the future is not fully determined by the past any more than the left-hand side of a picture is fully determined by its right-hand side. He discusses further the value of various contemporary developments in art. There is, in particular, an interesting description of his personal experience of seeing Rouault's painting *The Old Clown.*

May 22, 1960

Biederman first comments on the similarity of many of Bohm's views to those of Alfred Korzybski in *Science and Sanity*. He then raises questions about the instruments that serve as the perceiving intermediary between the physicist and nature: have these become too complicated? Has the physicist forgotten his or her own role in the process? Biederman thinks that Bohm's vision of totality is too extensive and suggests that instead one ought to focus on "certain arts and sciences". He emphasizes that man is now in charge of his own evolution and that a creative orientation is crucial. This is why art "holds the deepest significance to the future". He further discusses Bohm's views on the finite and the infinite, the past and the future, necessity and freedom, and asks for a more creative notion of determinism, something they go on to develop in future letters. He makes an interesting suggestion that " . . . the future is always in the process of implying . . . a newness to the past". There is finally a discussion of art in which Biederman openly reveals his personal opinions.

June 6, 1960

Bohm comments on the various books and articles Biederman has sent to him. In particular he expresses interest in Cézanne's ideas on space as a unity of interpenetrating planes and notes their similarity to some ideas which he is trying to develop on geometry. He is also in sympathy with Biederman's attempt to go beyond mimesis of nature while still trying to capture the general laws of process and relationships of nature. At the same time he questions whether Biederman is justified in claiming that Structurism is the main line in the evolution of art. He then goes on to clarify some of his previous ideas, emphasizing in particular that he does not want to reduce the large-scale world to the world of atoms, but that each level makes an irreducible contribution to the whole. For example, he suggests that the laws of living, thinking matter are not fully deducible from the laws of the constituent atoms. There is a further discussion of the limited and the unlimited and how the notion of a "thing" is to be understood. The discussion about the past and the future and about determinism vs. freedom also continues. Biederman has mentioned how Cézanne saw the universe as a pulsation of colour, and this prompts Bohm to explain his pulsating model of the electron, which emphasizes the indivisible unity of each thing with the cosmos and makes it possible, for example, to discuss the question of lawlessness in a new way.

June 28, 1960

Biederman begins by answering in the negative to Bohm's question about whether there is a dogmatism involved in seeing Structurism as the future of non-camera art. He admits that the structural view of nature is not readily accepted and discusses the various reasons for this. He then comments on Bohm's discussion of reductionism and the limitations of the micro-level, atomic description, and explains further the idea of "creative determinism", which becomes a key notion in their correspondence. He finally questions Bohm's way of regarding necessity and contingency as oppositions.

CHAPTER II: CREATIVE DETERMINATION

August 1, 1960

Bohm notes that Biederman's idea of "creative determinism" is close to his own idea of "the cosmic process forming itself and thus creating space, time, matter, movement, etc." Bohm prefers the term "creative termination" or "creative determination". There is a discussion of creativity in human relationships and a critique of the traditional idea of communication: ideas need to be produced "between" people, rather than being communicated from one person to others. This anticipates Bohm's later ideas on dialogue. It is noted that creativity in science and art may spread to human relations in general, giving rise to "a new mode of existence for humanity". He disagrees with Biederman on the question of "law and lawlessness, regularity and irregularity, necessity and contingency, etc.". Biederman appears to be claiming that things like lawlessness are subjective, arising out of our ignorance. We talk about "lawlessness" because we do not know the law in a given situation. Bohm claims that this denies reality to time, process and creativity itself. For him, a creative choice is not entirely determined by the past, although it may still be determined by something, namely the present moment itself. There is a discussion of pairs of concepts, such as actuality and possibility and necessity and contingency. Parts are by definition contingent, for the necessity of a part does not follow fully from what it is. On the other hand, "whatever the cosmos in its aspect of eternal totality is, it cannot be otherwise, for the simple reason that there is and can be only one such eternal totality". In other words, the whole cosmos is necessarily as it is. Bohm gives various arguments why contingency nevertheless is important, has an objective significance and is thus not merely the result of our ignorance. The letter ends with an interesting discussion of the question of possibility and actuality.

October 3, 1960

Biederman begins by commenting favourably on Bohm's general discussion of creativity and communication. Instead of the technologically coloured "communication" we need "conversation" between creative participants, in particular between man and nature. Both art and science are needed in determining the orientation of human life, for they have in them "qualities that could neutralize the destructive ones inherent in each field, since both have the potentiality for both good and evil". In reply to Bohm's criticisms, he says that abstraction is creative and time is not denied: "every creative act . . . is potentially open to new creation, creation itself is constantly undergoing change . . . This is based on an attitude of non-identity where anything is not the same as it was or will be or will ever be after that." He then claims that it is Bohm's notion of totality as a closed, necessary system which denies the creative and temporal quality. Another view of nature is possible, where the cosmic totality itself is seen as incomplete: "the cosmos itself is undergoing a constant state of growth, and is not any more 'self-limited' than any of the finite experiences we have". Biederman still does not see contingency as an objective feature but suggests that it is a feature of the incompleteness of our knowledge: "Contingency exists by virtue of the kind of structural abstracting relationship of the human being with nature . . . There would not be any contingencies if there were not humans to experience them." This makes clear that there is a tension between realism and anti-realism in Bohm's and Biederman's views on contingency and many other concepts. For Bohm contingency is, in a realist fashion, a feature of the mind-independent world, whereas for Biederman it is, in an anti-realist fashion, a feature dependent on human experiencing. Again with the notion of duality, Biederman argues that from the fact that we need to introduce dualisms in the analysis of nature it does not follow that there are dualisms in actuality. And similarly, with the notion of possibility: "possibility is only inherent in the interaction between a human being and nature, and not in extra-human nature as such". There is finally an illuminating discussion of the mimetic vs. creative artist. The mimetic artist primarily discerns the already present actualities perceivable in optical nature, whereas the "purely creative" or Structurist artist also takes optical perception further to focus on the process character of nature's structure: "the non-mimetic artist is a discoverer of hitherto non-actualized possibilities, which only human nature can actualize".

November 17, 1960

Bohm begins by discussing the question of the opposition of various categories and suggests that we view opposition from the point of view of process, which has one or more pairs of directions. A discussion of the universal and particular follows. In this context Bohm introduces notions such as "universalization" and "particularization", which clearly anticipate his later well-known notions of *unfolding–enfolding* and *explicate–implicate*. By discussing an example, Bohm shows how universalization and particularization are opposed to each other in direction but nevertheless, when seen in their totality, they are identical. He then discusses other oppositions, such as potential and actual, and necessity and contingency in a similar way. Biederman's idea of "creative abstractions" is commented upon in light of the above. There is also a critique of Biederman's idea that possibility is a subjective category. A discussion of "traces" again anticipates the implicate order: "A trace is . . . a structure in which the past is . . . folded . . . but in principle capable of being unfolded". Bohm compares the moments experienced in human consciousness with the moments of existence in nature. The rest of the letter then focuses on applying the ideas of necessitation and actualization of possibilities to understanding artistic creativity.

December 28, 1960

Biederman first compares physics and art: the physicists ponder minutely whatever they encounter, whereas artists, even Structurist ones, need not pursue their thought to the *n*th degree. He notes that aesthetics has tried to formulate a minute penetration of art experience but dismisses this as being too far removed from the actualities of art. There is then a commentary on Bohm's discussion of process with pairs of directions and a further discussion of whether necessitation occurs in reality or is a projection of our minds. The key idea is that because our views about necessity are likely to prove inaccurate in the future, it is only in a limited sense that our minds reflect what occurs in reality. Biederman underlines the difference in their approaches: "What you want to attribute to nature, I attribute exclusively to man." In response to Bohm's arguments that contingency is an objective feature of nature, Biederman concedes: "in some respect nature's objective reality could be otherwise where the intersection of man is possible". He then articulates his view of causality as "a process of diverse orders of creative determination". There is a further explanation of the nature of Structurist art and its relation to science: the Structurist artist "reveals those creations inherent in nature's process which man alone actualizes". Biederman finally comments carefully upon the various points Bohm has made about Structurism.

CHAPTER III: THOUGHT AND REALITY

February 2, 1961

Bohm expresses his astonishment about Biederman not admitting that a process of contingentation and necessitation goes on not only in art but also in nature. He then proceeds to describe his ideas on process. Drawing on the special theory of relativity, he discusses time and how the potentialities contained in an event are determined. He again tries to give a precise meaning to the idea that contingency is an objective category. The basically "inward" character of all process, whether natural or human, is emphasized in the sense that in each moment there is a trace of the rest of the world. He then moves on to discuss the role of past experience in determining our response to the world. In particular, he questions whether Biederman is limited by his evaluation of pre-Structurist art, and compares some of the Structurist ideas about evolution with Marxist ones.

March 2, 1961

Biederman again argues that contingency is unique to human life and that the more primitive levels of nature we examine, the more contingency gives way to necessity. It is human consciousness with its ability to choose which makes contingency prevalent. He then replies to Bohm's question about whether his evaluations of past art are too judgemental and thus not compatible with openness and creativity: "I try to comprehend all the major art attitudes prevailing, to literally project myself into the artist who made the art."

February 24, 1961

This letter breaks the "letter–reply" pattern that has been characteristic of the correspondence thus far, for Bohm writes another letter before receiving a reply. The order of letters in this volume are thus no longer in the chronological order based on when the letters are dated but rather, as much as possible, the order in which the ideas are discussed. Bohm focuses on the problem of contradiction. He starts off by referring to Biederman's point that our descriptions of reality are always limited and says that all of his own discussions attempt to "describe reality with the aim of definite *terms*, which may include words, symbols, pictures, etc.". This then leads on to a discussion of the relationship between thought and reality and also of *naming*, and of how thought works more generally – a description of cognition is given. He argues that contradiction and conflict are inevitable in the description of the world by words. We can, however, seek an "adequate reflection". Reality is seen as "implicit" in the sense that it cannot be understood solely by words. The rest of the letter contains, among many other interesting ideas, a careful discussion of the notions of difference and identity, a theme that is central in many of the later letters.

May 29, 1961

Biederman starts by discussing Bohm's idea that the attempt to describe reality with definite terms leads to contradiction, and suggests we begin with undefined terms instead. He also says that the notion of identity is "structurally destructive": "Differentiation, not identity, is the constant of experience." He finds it curious that Bohm places equal emphasis on "notions of identity" and "the differential aspects of process". He also suggests that we use the notion of *similarity* instead of identity to discuss the relation of the properties of objects. He then goes on to relate the rather abstract discussion he and Bohm have been having to his own work, especially by describing his experience when he first sees a piece of his own work. Because of the particular nature of his art, he sees it all at once: "Now, what I see during that first moment I will literally never see again in the actuality". There is also a fine description of the creative act, as experienced by Biederman himself: "I put *aside* all I know . . . At such times I have the experience of totality . . . because I have removed the barriers to being *in* it." He finally summarizes nicely the shift of focus of their discussion: "Until now we have been talking about the structure of our interpretations of reality, now we are talking about the structural functions of our talking."

CHAPTER IV: TRUTH AND UNDERSTANDING

December 22, 1961

After a long delay in the correspondence, Bohm begins by summarizing what to him seem the main recurrent themes. This list and discussion are very useful for the reader. There is: 1) the question of creative determination; 2) the problem of oppositions; 3) the problem of time; 4) the problem of totality; 5) our relationship to the world; 6) how to deal with the past creatively in science and art; 7) language, thought and semantics; 8) process; and 9) the role of abstraction. After summarizing these, Bohm moves on to discuss the notion of *understanding*, a topic that occupies him a great deal. What is basic to understanding is to see parts "as sides or aspects generated in a total process, so that you now understand why they are related as they actually are". He further emphasizes that thought, feeling and action are sides of a single process. Truth is an adequate understanding of each moment rather than something fixed, absolute and everlasting. In the latter part of the letter Bohm tries to characterize the notion of identity with the help of the notion of the field: "two processes, A and B, are identical in a certain field, if they differ by the nullity of that field, i.e., by something which produces no effects in the field in question". He is trying to save the content of the concept of identity so that we can justify the common usage of words.

December 26, 1961

In this letter, written only four days after his previous one, Bohm further describes his reactions to Structurist art. Having noted that a great deal of the work strikes him as cold and calculated, he asks whether it is necessary for Structurist art to restrict itself to planes: "Doesn't nature structure have curves in it too?" He then applies the concept of contingentation and necessitation from earlier letters to Carl Visser's art. He moves on to discuss randomness, and by considering an example of throwing coins, he tries to demonstrate that "relative randomness" has a clear and well-defined meaning. There is an interesting discussion of a random relationship involving a kind of independence of two orders. It is worth considering this discussion of the concept of order in relation to Bohm's later ideas of the implicate and explicate order. At the end of the letter Bohm emphasizes that creativity cannot be planned or calculated by a mathematical formula – a theme similar to those much discussed today, e.g. by Roger Penrose.

January 18, 1962

Biederman refers to three letters, of Dec. 22 and 26, 1961 and Jan. 13, 1962. The Jan. 13 is (probably a short) airmail letter which has not been preserved. Biederman says in the beginning that he is only replying to the Dec. 22 letter, but in fact there is also a reply to the Dec. 26 letter later on. In replying to Bohm's Dec. 22, 1961 Biederman first defends his idea of using the concept of similarity instead of identity. He then agrees with Bohm's view of creation as "instantaneous thought–feeling–action" as well as with the idea that true creativity involves the absence of the "I" or Ego. Bohm has said that the understanding of totality goes beyond any of the particular senses. For Biederman this amounts to giving understanding a transcendental status over the senses. He sees this as a bias on the side of the physicist, who has to picture himself in a world devoid of human habitation. He then comments on Bohm's ideas on the relationship between understanding and the feeling of love. Biederman's reply to Bohm's Dec. 26, 1961 tackles Bohm's question about why the Structurist does not use curves but sticks to straight planes: "Since the Structurist seeks the example of nature for the structural possibilities of his developments, the cube puts him in a position to begin on the simplest structural terms possible." Biederman gives further replies to Bohm's questions about his art and comments somewhat critically on Bohm's discussion of randomness. Finally Biederman expresses his fascination concerning Bohm's remarks on mathematics as creation, and considers the relation between art, music and mathematics.

December 29, 1961

The intention of this letter is to summarize and add a few points to Bohm's previous two letters (Dec. 22 and Dec. 26, 1961). Bohm has not yet received Biederman's reply to those letters (Jan. 18, 1962) and thus here comments on Biederman's letter of May 29, 1961. He again emphasizes, unlike Biederman and Korzybski, that the notion of similarity is not enough, but a stronger notion of identity or sameness is required, if language is to function. For example, if two types of red, R_1 and R_2, are not the same in some respect, "there would be no justification for using the same word (red), and language would turn into utter chaos". The letter further discusses topics like "projecting the world into fields", "the concrete and the abstract" and "reflection". Bohm then returns to the notions of understanding and truth: "Truth is not just something wholly outside ourselves, which we approach step by step; nor is it wholly internal and subjective. Rather, it is something that comes into being within us, by which we are related to totality." He finally restates his view that contingency (just like necessity) is an essential part of the relationships in "nature structure".

December 30, 1961

Bohm comments further on Biederman's letter of May 29, 1961. He agrees, generally speaking, that we should begin with undefined terms and not with definite terms, but notes that there is a deeper problem of why one can use any terms at all, defined or undefined, and goes on to discuss how to tackle the problem. Towards the end of the letter Bohm considers the complex notion of "difference between similarity and difference" and notes that this is fundamental to our understanding of anything whatsoever.

February 24, 1962

Biederman admits that Bohm's last two letters (presumably Dec. 29 and 30, 1961) have moved him towards Bohm's point regarding the implicit character of identity. Biederman, however, still resists the term "identity". He characteristically suggests that identity is constructed by the human mind: "identity exists only when the individual deliberately or subconsciously identifies". He still sees identity as alien to his approach which emphasizes change, growth and movement: "however thin you slice the notion of identity with quotes, it has only the verbal function of frustrating the actualities of process orientation". He notes that Bohm's remarks on mathematics, art and music are interesting and adds that "in art the eye does not see mathematically". He also agrees with some of Bohm's comments on Structurist art. For example, mimeticism is not cancelled by Structurist art but is "a unique form of visual experience that is far from exhausted" and which continues an evolution via camera art in all its forms. The uniqueness of Structurist art lies in that it "does not, in any way, reflect what nature has created. Rather, it is a direct reflection of the total creative process of nature. Consequently, *entirely* unique events arise with this form of art in its amalgamation of man with nature." At the end of the letter Biederman returns to their ongoing debate about whether contingency is an objective aspect of nature. While agreeing that Bohm has presented some good arguments, Biederman remains dissatisfied: "If you as a human regard contingency as ever present in all your abstractions from nature, how do you know you are not projecting contingency upon nature, rather than reflecting it?"

CHAPTER V: BEYOND THE SUBJECT–OBJECT DISTINCTION

February 2, 1962

Bohm begins by underlining the importance of the problem of understanding. He notes that the mind functions in two ways: first through associative thinking plus the drafting of conclusions by logical arguments and calculations and second, via understanding, the perception of totality. Understanding itself is ultimately undefinable; the action of understanding is not verbalizable, but it can be initiated in the mind of another by the proper use of words.

February 9, 1962

This is a short letter in which Bohm again expresses his dissatisfaction with the notion of similarity: because in some way everything is similar to everything else, it is necessary to consider something like "essential similarity" and to understand a basic concept such as "democracy" in its essential totality.

March 17, 1962

Bohm acknowledges receipt of a letter from Biederman, most probably the letter dated February 29, 1962. He then comments on the photographs of Biederman's work, appreciating in particular Biederman's use of colour. There is an approval of Biederman's point that even in mimetic art the content is not imposed from nature but is also the result of the artist's meeting with nature. Bohm then questions whether photography need be the only line of development for mimesis. He also formulates what he sees Biederman's art to be doing; Structurist art is not just a reflection of nature nor even of man's meeting with nature: "Rather, it should be a unique form of reality in itself, including a reflection both of nature's creative process and of man's peculiar and unique contribution to this process, but also including still more – namely – itself." He further discusses the relation between mathematics and art: beauty has a role in science, just as every beautiful work of art must be "true". More importantly, the scientist, the mathematician and the artist will all benefit from a better understanding of other fields. For example, it is important for the scientist to understand his or her own level of experience because all data finally come from perceptions and abstractions at this level. The artist is helpful in providing a new way of understanding ordinary perceptual experience. In order to explore the notion of similarity, Bohm considers similar but different forms of democracy as an example. This leads to a lengthy discussion of democracy and other political views. Importantly, Bohm is willing to replace "sameness" by "similarity", in other words to drop "identity" from their set of concepts, as long as certain conditions are accepted. This is a vivid example of how the ideas of the two men change and develop in the course of the correspondence. Finally Bohm proposes an argument for Biederman to think over, an argument that says that difference is the logically prior category and similarity is a special kind of difference.

March 18, 1962

This is a short letter in which Biederman comments favourably on Bohm's essay "On the Problem of Truth and Understanding in Science" (1964): "If the 'understanding' I experienced in your essay means anything, then assuredly you are on the track of something tremendous."

April 14, 1962

Biederman comments further on Bohm's essay on "Truth and Understanding", as well as the letters of March 17 and 23, 1962. He is impressed by some of Popper's ideas on falsifiability (in particular the idea that there is no final verification). He also notes that some kind of subjectivism is inescapable, and that it also makes room for creativity: the expression of man's subjectivism achieves, as in Structurism, what would never appear without man in nature. He finally approves of Bohm's notion that truth is not static but in a state of constant change.

April 23, 1962

Biederman begins by replying to Bohm's question of whether photography will be the main line of development for mimeticism. For Biederman the question is whether new mimetic ways are possible, besides photography, that will sustain the evolution of art. The key point for both human life and the development of art is to expand man's experience of nature. Biederman thinks that most post-Courbet mimetic art has worked to make the subjective independent of objective nature; this is why he sees only certain camera-artists as succeeding in taking mimesis further. He comments very favourably on Bohm's attempt to summarize Structurist art: "You are the first person to write to me who understands." He disagrees only with Bohm's point that Structurism is not "even man's meeting with nature", and says rather that it is "man's meeting with nature ever anew". He further comments on Bohm's discussion of the relationship between science, art and mathematics. There is a discussion of the question of democracy and other political views, and Biederman is in general in agreement with Bohm. Finally, Biederman agrees with Bohm on the idea that similarity is a special kind of difference.

March 23, 1962

Bohm continues discussing the idea that similarity is a kind of difference by suggesting that there are two kinds of similarity. First, A and B can be similar in the field of C, which means that in this field their difference does not show. Second, he proposes the idea of "similarity for self": each self determines a natural field of abstraction in which there is essential difference and essential similarity. He then moves on to discuss the action of understanding, the act of perceiving a totality, both inwardly and outwardly. To divide experience into the inner and the outer is ultimately a habit; the division is done for convenience and for security, but it also causes many problems. It is thus crucially important to understand the underlying unity of the inner and the outer, the self and the non-self. Bohm also describes his own experiences when trying to see this unity: "I occasionally got a sudden 'glimpse' in which one felt that reality is in a different dimension . . . one saw that the inner and outer are basically one. . . . In this state of unity . . . the new truth starts to operate." Finally love and understanding are both required to maintain this unity. The above ideas probably reflect Bohm's relatively recent interest in Krishnamurti's approach, although no explicit reference to Krishnamurti is yet made. (In the letters written after the period published in this volume, Krishnamurti soon becomes a central figure. One of the reasons why the correspondence ended in 1969 was precisely disagreement about the importance of Krishnamurti. For Bohm he was crucial while Biederman remained unconvinced about his importance.) Bohm then discusses again his experience when seeing Rouault's painting *The Old Clown*, now in relation to the problem of the "inner" and the "outer". This painting, he says, is an example of how non-camera mimesis could still be useful. He suggests that Structurist art ought to and could explore the unity of the subject and the object; in quantum physics this has already become central. Finally, there is a discussion of the role of pre-verbal experience and abstraction.

April 27, 1962

Biederman first addresses Bohm's idea that there are two kinds of similarity. In particular, Biederman criticizes Bohm's use of the term "self", suspecting that Bohm's use is "flavored with some identity". He then comments on Bohm's discussion of the act of understanding, relating it to his own ideas on art: "Every act of understanding is unique and can only be repeated by recollection and never again by direct perception." Bohm's discussion of the inner and the outer is "a refreshing outlook on an ancient problem". Biederman regrets the split between man and nature which is now sharper than ever. He is particularly fascinated by Bohm's description of a glimpse of another dimension: "The critical character of the glimpses you describe is that they are non-verbal, the difference between cogitation about and the experience of being totality". There is a discussion of love and understanding, in which Biederman feels that Bohm's ideas are too much directed towards an ideal goal. He then discusses Bohm's experience with the Rouault painting. Not denying the experience itself, Biederman remains critical of Rouault's Expressionism. He then discusses the way in which Structurist art addresses the question of the inner and the outer: "the work must indeed radiate a unity with surrounding nature". Finally there is a discussion of the importance of pre-verbal experience.

April 15, 1962

Bohm makes a systematic list of the topics he wishes to cover in response to Biederman's letters of January 18 and February 24, 1962. This letter acts as a kind of summary and is very useful to the reader. Bohm's list of topics includes: 1) creation in art, music and mathematics; 2) the Ego; 3) contradiction in thinking; 4) understanding and the understood; and 5) randomness, chance and contingency, etc. The letter discusses all these points. For example, Bohm argues that there is no contradiction between randomness on the small scale and statistical regularity on the large.

May 7, 1962

Biederman replies to the various points taken up by Bohm in his April 15, 1962 letter. He reports that he has tried to experiment with going beyond the subject–object distinction in the way Bohm has suggested. He still remains dissatisfied with the notion of randomness and disorder: "Until the disorder view is proved, the order view must reside at least with equal attention in our perceptive activities."

April 24, 1962

The letter begins by addressing Biederman's question about the necessity of having "a spectrum of descriptions of experience", e.g. "thinking–feeling and feeling–thinking". Bohm develops this into what he calls "thought–feeling–action" and emphasizes that action is the most fundamental, including the act of understanding. The letter ends by underlining a point which has by now become a key one in the correspondence: "In this point of view, the separation of 'subject' and 'object' is dropped. What is fundamental is the act or process itself."

BIBLIOGRAPHY

BOOKS BY DAVID BOHM

The Quantum Theory, Prentice-Hall, Englewood Cliffs, New Jersey (1951).

Causality and Chance in Modern Physics, Routledge & Kegan Paul, London ([1957] 1984).

The Special Theory of Relativity, The Benjamin/Cummings Publishing Company, Inc., Advanced Book Program, Reading, Massachusetts (1966).

Fragmentation and Wholeness, The van Leer Jerusalem Foundation, Jerusalem (1976).

Wholeness and the Implicate Order, Routledge & Kegan Paul, London (1980).

Unfolding Meaning: A Weekend of Dialogue with David Bohm, ed. Donald Factor, Foundation House Publications, Mickleton (1985).

Co-authored with J. Krishnamurti, *The Ending of Time*, Harper & Row, San Francisco (1985).

Co-authored with J. Krishnamurti, *The Future of Humanity*, Mirananda, The Hague (1986).

Co-authored with D. F. Peat, *Science, Order and Creativity*, Bantam, New York (1987).

Co-authored with Mark Edwards, *Changing Consciousness: Exploring the Hidden Source of the Social, Political and Environmental Crises Facing Our World*, Harper, San Francisco (1991).

Co-authored with B. J. Hiley, *The Undivided Universe: An Ontological Interpretation of Quantum Mechanics*, Routledge, London (1993).

Thought as a System, Routledge, London (1994).

On Dialogue, Routledge, London (1996).

On Creativity, Routledge, London (1998).

SELECTED ARTICLES BY DAVID BOHM

"A Suggested Interpretation of the Quantum Theory in Terms of Hidden Variables I & II", *Phys. Rev.* vol., 85, no. 2, pp. 166–193 (1952). Republished in J. A. Wheeler and W. H. Zurek (eds) *Quantum Theory and Measurement*, 369–96, Princeton University Press, Princeton (1983), pp. 369–96.

"On the Relationship between Methodology in Scientific Research and the Content of Scientific Knowledge", *The British Journal for the Philosophy of Science*, vol. XII (1961), pp. 103–16.

"Classical and Non-Classical Concepts in the Quantum Theory", *The British Journal for the Philosophy of Science*, vol. XII, no. 48 (1962), pp. 265–80.

"Problems in the Basic Concepts of Physics" (an inaugural lecture delivered at Birkbeck College, February 1963). In *Satyendranath Bose 70th Birthday Commemoration Volume*, Part II, Calcutta (1965), pp. 279–318.

"On the Problem of Truth and Understanding in Science", in M. Bunge (ed.) *The Critical Approach to Science and Philosophy, in Honor of Karl R. Popper*, Collier Macmillan, London (1964), pp. 212–23.

"A Proposed Topological Formulation of the Quantum Theory", in I. J. Good (ed.) *The Scientist Speculates*, Putnam, New York (1965), pp. 302–14.

"Space, Time and the Quantum Theory Understood in Terms of Discrete Process", *Proceedings of the International Conference on Elementary Particles*, Kyoto (1965), pp. 252–86.

"On Creativity", *Leonardo*, vol. 1 (1968), pp. 137–49.

"On the Relationship of Science and Art", in A. Hill (ed.) *Data: Directions in Art*, Faber & Faber, London (1968).

"Some Remarks on the Notion of Order and Further Remarks on Order", in C. H. Waddington (ed.) *Towards a Theoretical Biology*, vol. 2, Aldine Press, Chicago (1970), pp. 18–40.

"On Bohr's Views Concerning Quantum Theory", in T. Bastin (ed.) *Quantum Theory and Beyond*, Cambridge University Press, Cambridge (1971), pp. 33–40.

"Science as Perception–Communication", in F. Suppe (ed.) *The Structure of Scientific Theories*, University of Illinois Press, Illinois (1977), pp. 374–423.

"Time, the Implicate Order and Pre-Space", in D. R. Griffin (ed.) *Physics and the Ultimate Significance of Time*, State University of New York Press, New York (1986), pp. 172–208.

"Meaning and Information", in P. Pylkkänen (ed.) *The Search for Meaning, The New Spirit in Science and Philosophy*, Thorsons Publishing Group, Wellingborough (1989), pp. 43–62.

"A New Theory of the Relationship of Mind and Matter", *Philosophical Psychology*, no. 3 (1990), pp. 271–86.

SELECTED BIBLIOGRAPHY OF BOOKS AND ARTICLES ON DAVID BOHM

Cushing, J. T., Fine, A. and Goldstein, S. (eds), *Bohemian Mechanics and Quantum Theory*, Kluwer, Dordrecht (1996).

Griffin, D. (ed.), *Physics and the Ultimate Significance of Time: Bohm, Prigogine and Process Philosophy*, State University of New York Press, New York (1986).

Hiley, B. J. and Peat, F. D. (eds), *Quantum Implications: Essays in Honour of David Bohm*, Routledge, London (1987).

Peat, F. D., *Infinite Potential: The Life and Times of David Bohm*, Addison Wesley Publishing Company Inc., Reading, Mass. and Harlow, England (1996).

Pylkkänen, P. (ed.), *The Search for Meaning*, Thorsons, Wellingborough (1989).

Schindler, D. L. (ed.), *Beyond Mechanism*, University Press of America, Lanham (1986).

BOOKS BY CHARLES BIEDERMAN

Art as the Evolution of Visual Knowledge, Red Wing, Minnesota (1948).

Letters on the New Art, Red Wing, Minnesota (1951).

The New Cézanne: from Monet to Mondrian, Red Wing, Minnesota (1958).

Search for New Art, Red Wing, Minnesota (1979).

Art, Science, Reality, Red Wing, Minnesota (1988).

The Dehumanization and Denaturalization of Modern Art, Red Wing, Minnesota (1992).

Nature and Art Anew, Red Wing, Minnesota (1993).

The End of Modernism: Figurative or Abstract, Red Wing, Minnesota (1994).

SELECTED ARTICLES BY CHARLES BIEDERMAN

"Art and Science as Creation", *Structure*, vol. I, no. 1, Amsterdam (1958), pp. 2–18.

"Instinct–Intuition and Emotion–Intellect in Art", *The Structurist*, no. 1, Saskatoon (1960–1), pp. 42–51.

"A Non-Aristotelian Creative Reality", *Structure*, vol. IV, no. 2, Amsterdam (1962), pp. 38–43.

"Art in Crisis", *Studies in the Twentieth Century*, no. 1, Troy, NY (1968), pp. 39–59.

"Dialogue II: Creative or Conditioned Vision", *Data: Directions in Art, Theory and Aesthetics*, London (1968), pp. 76–94.

SELECTED BIBLIOGRAPHY OF ARTICLES ON CHARLES BIEDERMAN

Craven, D., "The Art of Charles Biederman", unpublished PhD thesis, University of North Carolina, Chapel Hill (1980a).

Craven, D., "Charles Biederman's Art of Complicity", *Arts Magazine*, March (1980b), pp. 130–5.

Denny, Robyn, "Introduction", in *Charles Biederman: A Retrospective Exhibition with Especial Emphasis on the Structurist Works of 1936–69* (exhibition catalogue for the Hayward Gallery, London, and the Museum and Art Gallery, Leicester), Arts Council of Great Britain, London (1969).

Hill, A., "The Climate of Biederman", *Studio International*, vol. 178, no. 914, London (1969), pp. 69–70.

Kuspit, D. B., "Charles Biederman's Abstract Analogues for Nature", *Art in America*, May/June (1977), pp. 80–3.

Van der Marck, J., "Biederman and the Structurist Direction in Art", in *Charles Biederman: A Retrospective Exhibition with Especial Emphasis on the Structurist Works of 1936–69* (exhibition catalogue for the Hayward Gallery, London, and the Museum and Art Gallery, Leicester), Arts Council of Great Britain, London (1969).

INDEX